International Marine Science Affairs

NATIONAL ACADEMY OF SCIENCES

International Marine Science Affairs

A Report by the
International Marine Science Affairs Panel
of the
Committee on Oceanography

NATIONAL ACADEMY OF SCIENCES
Washington, D.C. 1972

ISBN 0-309-01937-0

Library of Congress Catalog Card Number 74-183584

Available from
Printing and Publishing Office
National Academy of Sciences
2101 Constitution Avenue, N.W.
Washington, D.C. 20418

Printed in the United States of America

Preface

Soon after its formation early in 1968, the International Marine
Science Affairs Panel (now the Committee on International Marine
Science Affairs Policy) recognized the need to develop a better understanding of the requirements for international cooperative action
and of the mechanisms needed for such action in order that the
Panel could respond to agencies requesting advice in this area on
short notice. For the next two years, the Committee devoted a main
portion of its efforts to developing such understanding. The report
is the result of its deliberations during that interval.

In a long-term project of this kind, events can move more quickly
than the study. Thus, some of the recommendations put forth in this
report have already been partially implemented or are awaiting official action. Where possible, this is noted in the text or in footnotes.

From the beginning of the study to six months before its completion, the Committee benefited from the wisdom and experience of
its late colleagues, Dr. Wilbert M. Chapman and Dr. Milner B. Schaefer.
Each had a major part in determining the overall structure and much
of the detailed content of this report.

While the entire Committee participated in the numerous discussions from which the report evolved, drafting chores were under-

taken by F. G. Blake, W. Burke, D. Cheever, H. Kasahara, D. Leipper, G. Pontecorvo, and W. Wooster. Final revisions were made by W. Burke, H. Kasahara, J. Knauss, and W. Wooster. The Committee is grateful to A. Denis Clift for his help in drafting the chapter on global services and to E. Roy Dillon, then Acting Executive Secretary of the National Council on Marine Resources and Engineering Development, for his helpful cooperation.

Although this report does not form a part of the program of either institution, the Committee is grateful for staff assistance and support from the Sea Grant Program of the Division of Marine Resources, University of Washington, and from the Center for Marine Affairs, Scripps Institution of Oceanography. The Office of the Foreign Secretary of the National Academy of Sciences furnished aid in support of preparation of this report. Mrs. Joyce Weinberger of the secretarial staff of the University of Washington School of Law provided expert assistance in facilitating completion of the Committee study.

William Burke, *Chairman*
International Marine Science
Affairs Policy Committee

International Marine Science Affairs Panel

William T. Burke, *Chairman*
  *University of Washington Law
    School
  Seattle, Washington*

F. Gilman Blake
  *Senior Research Scientist
  Chevron Research Corporation
  La Habra, California*

Harrison Brown
  *Foreign Secretary
  National Academy of Sciences
  Washington, D.C.*

John C. Calhoun, Jr., (*ex officio*)
  *Vice President for Programs
  Texas A&M University
  College Station, Texas*

*Wilbert M. Chapman
  *Director, Marine Resources
  Ralston Purina Company
  San Diego, California*

Daniel Cheever
  *Professor, Graduate School of
    Public and International Affairs
  University of Pittsburgh
  Pittsburgh, Pennsylvania*

Sigfried Ciriacy-Wantrup
  *Professor Agricultural Economics
  University of California, Berkeley
  Berkeley, California*

Hiroshi Kasahara
  *Associate Dean
  College of Fisheries
  University of Washington
  Seattle, Washington*

Dale F. Leipper
  *Department of Oceanography
  Naval Postgraduate School
  Monterey, California*

*Deceased.

v

Arthur H. Maxwell
  Associate Director
  Woods Hole Oceanographic
    Institution
  Woods Hole, Massachusetts

Giulio Pontecorvo
  Graduate School of Business
  Columbia University
  New York, New York

Roger Revelle (corresponding
  consultant)
  Harvard Center for Population
    Studies
  Harvard University
  Cambridge, Massachusetts

*Milner B. Schaefer
  Director, Institute of Marine
    Resources
  University of California, San Diego
  La Jolla, California

William L. Sullivan, Jr. (consultant)
  Chief, Oceanography and Interna-
    tional Organizations
  Department of State
  Washington, D.C.

Warren S. Wooster
  Scripps Institution of Oceanography
  La Jolla, California

*Deceased.

Ocean Affairs Board

ACRONYMS USED IN THE REPORT

ACC	Administrative Coordination Committee (UN)
ACMRR	Advisory Committee on Marine Resources Research (FAO)
ACOMR	Advisory Committee on Oceanic Meteorological Research (WMO)
AEC	Atomic Energy Commission
AGU	American Geophysical Union
AFS	American Fisheries Society
ASLO	American Society of Limnology and Oceanography
BCF	Bureau of Commercial Fisheries (now NMFS)
CICAR	Cooperative Investigation of the Caribbean and Adjacent Regions
CINECA	Cooperative Investigation of the Northern Part of the Eastern Central Atlantic
CIOA	Committee on International Ocean Affairs
CIPME*	Committee on International Policy in the Marine Environment (now CIOA)
CMG	Commission on Marine Geology (IUGS)
COMSER	Commission on Marine Science, Engineering and Resources (Stratton Commission)
COFI	Committee on Fisheries (FAO)
COLD	Council of Oceanographic Laboratory Directors
CSK	Cooperative Study of the Kuroshio (IOC)

*Organization acronyms that are no longer current.

DNP	Declared National Program
DOD	Department of Defense
EASTROPAC	An Oceanographic study of the Eastern Tropical Pacific
ECOR	Engineering Committee on Oceanic Research
ECOSOC	Economic and Social Council of the United Nations
EPA	Environmental Protection Agency
ESSA*	Environmental Science Services Administration (now NOAA)
FAO	Food and Agricultural Organization of the United Nations
GA	General Assembly of the United Nations
GESAMP	Group of Experts on the Scientific Aspects of Marine Pollution
HEW	Health, Education and Welfare, Department of
IABO	International Association of Biological Oceanography (IUBS)
IAEA	International Atomic Energy Agency (UN)
IAPSO	International Association for the Physical Sciences of the Ocean (IUGG)
IATTC	Inter-American Tropical Tuna Commission
IBP	International Biological Program (ICSU)
IBRD	International Bank for Reconstruction and Development (UN)
ICES	International Council for the Exploration of the Sea
ICITA	International Cooperative Investigation of the Tropical Atlantic
ICJ	International Court of Justice
ICMSE	Interagency Committee on Marine Science and Engineering
ICNAF	International Convention for the Northwest Atlantic Fisheries
ICO*	Interagency Committee on Oceanography
ICSEM	International Commission for the Scientific Exploration of the Mediterranean
ICSPRO	Inter-Secretariat Committee on Scientific Programmes Related to Oceanography
ICSU	International Council of Scientific Unions
IDOE	International Decade of Ocean Exploration
IGOSS	Integrated Global Ocean Station System (IOC)
IGY*	International Geophysical Year (ICSU)
IHB*	International Hydrographic Bureau (now IHO)
IHB	International Hydrographic Bureau (Executive Board of IHO)
IHO	International Hydrographic Organization
IIOE	International Indian Ocean Expedition (SCOR)
ILC	International Law Commission
IMCO	Intergovernmental Maritime Consultative Organization (UN)
IMSAP	International Marine Science Affairs Panel (now the Committee on International Marine Science Affairs Policy)
IOC	Intergovernmental Oceanographic Commission (UNESCO)
IPHC	International Pacific Halibut Commission

*Organization acronyms that are no longer current.

IPSFC	International Pacific Salmon Fisheries Commission
ITU	International Telecommunication Union (UN)
IUBS	International Union of Biological Sciences (ICSU)
IUGG	International Union of Geodesy and Geophysics (ICSU)
IUGS	International Union of Geological Sciences (ICSU)
LEPOR	Long-Term and Expanded Program of Oceanic Exploration and Research (IOC)
MTS	Marine Technology Society
NAS	National Academy of Sciences
NASCO*	National Academy of Sciences Committee on Oceanography
NATO	North Atlantic Treaty Organization
NCMRED*	National Council of Marine Resources and Engineering Development
NMFS	National Marine Fisheries Service of NOAA
NOAA	National Oceanic and Atmospheric Administration
NSF	National Science Foundation
OAS	Organization of American States
ODAS	Ocean Data Acquisition Systems
OECD	Organization for Economic Cooperation and Development
ONR	Office of Naval Research (DOD)
PIPICO	Panel on International Programs and International Cooperation in Ocean Affairs (formerly "in Oceanography")
PSMSL	The Permanent Service for Mean Sea Level
RVOC	Research Vessel Operators Council
SCAR	Scientific Committee on Antarctic Research (ICSU)
SCIBP	Special Committee on the International Biological Program (ICSU)
SCOPE	Special Committee on Problems of the Environment (ICSU)
SCOR	Scientific Committee on Oceanic Research (ICSU)
UN	United Nations
UNDP	United Nations Development Program (UN)
UNESCO	United Nations Educational, Scientific and Cultural Organization (UN)
WDC	World Data Center
WHO	World Health Organization
WMO	World Meteorological Organization

*Organization acronyms that are no longer current.

Contents

CHAPTER 1

Introduction and Summary of Recommendations

INTRODUCTION

Recent studies of international marine affairs by numerous private and governmental bodies emphasize that significant progress in understanding the ocean, the processes operating therein, and its resources requires the cooperation of many nations, developing as well as developed. Seldom, however, are these general injunctions of international cooperation accompanied by specific indication of necessary future actions. In the light of the marked increase in the intensity and variety of ocean uses and of the consequent heightened stress and political conflict as institutions of governance adapt to new requirements for maintaining public order, a specific outline for future international actions is needed.

In this study, the International Marine Science Affairs Panel has outlined some proposed actions. The development of the study was threefold. First, the important tasks pertaining to marine science that demand international action for their accomplishment were identified. Second, the international mechanisms available for taking the desired international action and the national mechanisms that activate them were determined. The final step was to consider the adequacy of the available mechanisms and to make recommendations for their improvement.

1

In carrying out analysis on these three levels, the report briefly describes existing mechanisms for international action and comments about some of the factors accounting for them. In the light both of this description and explanation and of likely future developments, recommendations are advanced for new or improved institutions (national and international) and better substantive programs. These recommendations should lead to an improvement in the state of international marine science affairs.

Three major considerations underlie the need for improvement of international institutions and programs concerning the marine environment: (1) for all the vast size, complexity, richness in resources, and grandeur of the world ocean, man's involvement with it intimately and profoundly reflects the growing interdependencies of a world organized into a growing number of separate political units; (2) there is increasing awareness that interdependencies require deference to and regard for others by conscious collaboration in many forms; and (3) the increasingly rapid progress in development of marine sciences and technology is laying the base for intensified, multiple uses of the ocean and its resources. A consequence of these concurrent developments is widespread public and official interest in the means for subjecting ocean activities to a political authority capable of coping with the effects of pronounced interdependencies and intensified multiple uses. As man's perspectives change in response to these new perceptions and understandings, he recognizes the need to be concerned about the collective impact of many separate activities in the ocean. The concern centers on political institutions for dealing with this collective impact in order to achieve the common objective: an effective utilization of the marine environment that provides both for its preservation for later generations and for an equitable sharing of growing benefits among the current generation.

There are numerous national and international political institutions (both public and private) now attempting to contribute to this objective. The purpose of this report is to examine these institutions in terms of their usefulness to the goals of wise, effective, and equitable ocean use. Recommendations are intended to suggest a means to obtain these goals.

The remainder of this introduction offers a summary statement of the three-step analysis. The body of the report is composed of five chapters, each of which discusses in greater detail one of the tasks requiring international action, identifies mechanisms for action, and offers recommendations. Most of the recommendations are directed to various agencies of the United States Government.

We find that international cooperation is needed in marine science principally to cope with five major tasks:

- The study of ocean processes
- Provision of global marine science services
- Regulation for rational use of the ocean
- Facilitation of ocean research activities
- Assistance to developing nations

Significant gains in understanding in two general areas of ocean research can be achieved through positive international cooperation. Investigations of processes that are of ocean dimensions can be advanced through such cooperation because of the vast scale involved and the relative slowness with which observations can be made. In applications-oriented research, international cooperation is essential to establish an agreed basis for rational use of the ocean and its resources both nationally and internationally.

As ocean uses intensify on a global scale, there is a growing need for provision of global marine science services. Among the services requiring international action are

- Exchange of information
- Standardization and intercalibration of methods
- Cooperative investigations and assessments
- Technical services such as navigation aids, allocation and management of radio frequencies, and forecasting and warning services
- Monitoring and surveillance

Regulation for rational use of the ocean necessarily entails international cooperative action because governance of marine affairs is still largely conducted through nation-states whose boundaries extend into and divide the ocean. While most of the ocean lies beyond national boundaries, some international uses of the ocean necessitate crossing national boundaries in the ocean. Recently, ocean use has both diversified and intensified, with new forms of use being initiated and traditional practices, such as fishing, being increased. This recent development, occurring mostly since 1955, reflects an increasing demand for some ocean resources, due in part to the expansion of population and the press of scientific discovery and technological innovation. The nature of modern military use of the ocean differs from the historical pattern, in that naval forces now have enormous strategic importance as part of a larger inventory of weapons systems. These various factors combine to produce political conflict and controversy that only international cooperative action can resolve without violence. Because of the diversification of ocean uses, one use of the ocean sometimes interferes with another. In particular, using the ocean as a receiver for wastes is often incompatible with fishing, recreation, and other uses. Both the scientific understanding and regulation of conflicting uses call for international action.

A fourth major task is to facilitate scientific investigation in the ocean by eliminating or minimizing legal impediments to research and by protecting the instrumentalities placed in the ocean to gather data. Coastal states are increasingly interfering with research by extending their laws to forbid or condition investigations in adjacent regions. These restraints are now widely recognized, and it is increasingly apparent that the principal means for coping with them successfully is international cooperative activity.

The fifth task is to provide assistance to states desiring either to develop a stronger marine science base or to make more effective use of marine resources for development and other purposes. Obviously, international action is inherent in such technical assistance, but international understanding and cooperation are also needed to cope with the many problems that arise in the provision of this assistance through several forms international action might take.

MECHANISMS FOR INTERNATIONAL ACTION

A great variety of mechanisms are available for pursuing international action to help in accomplishing the foregoing tasks. On the public (governmental) international level, the major mechanisms are the United Nations and the several specialized agencies that undertake activities relating to the marine environment, including the Food and Agriculture Organization of the United Nations; the United Nations Educational, Scientific and Cultural Organization; the World Meteorological Organization, and the Intergovernmental Maritime Consultative Organization. Occasionally special conferences are convened by the United Nations or one of the agencies to deal with particular problem areas. Outside the UN system, there are numerous bodies that operate on a regional or local basis, particularly the several international fisheries commissions.

There are also numerous international private bodies specializing in marine affairs that provide advice to public bodies. Several of these are in the International Council of Scientific Unions, the principal one being the Scientific Committee on Oceanic Research, which is a major source of external scientific advice to the Intergovernmental Oceanographic Commission of UNESCO and to UNESCO. Other component bodies of ICSU are especially important as mechanisms for communications among scientists.

In addition to international organizations, there are less structured mechanisms available to all nations. States commonly interact by concluding *ad hoc* bilateral and multilateral agreements that regulate their conduct vis-à-vis each other. Unilateral action by states (and the reciprocal behavior or acquiescence of other states) was historically an important means for maintaining order, and this method is still employed today. The vastly greater number of states relative to earlier times and the immensely more complicated issues in dispute bring the usefulness of unilateral action into question.

National mechanisms within the United States are important to the present study as means for generating or implementing international action. The recently created National Oceanic and Atmospheric Administration is an important new domestic mechanism that is highly significant for U.S. participation in international mechanisms. At this writing, however, responsibility for U.S. ocean foreign policy is still widely diffused among several offices in the Department of State.

RECOMMENDATIONS

After surveying the tasks requiring international action and assessing the mechanisms available, the Panel reached a consensus on a number of recommendations. The major recommendations are briefly summarized here and elaborated in the individual chapters.

INTERNATIONAL SOCIAL STUDIES (p. 11)

International cooperation in social studies on appropriate aspects of the use of the ocean and its resources is needed and should be encouraged.

INTERNATIONAL ROLE OF NOAA (p. 22)

The National Oceanic and Atmospheric Administration should serve as the focal point for U.S. participation in international ocean investigations. An office within NOAA should keep under review all documents and reports on relevant scientific and engineering matters dealt with by IOC, WMO, FAO, IMCO and other intergovernmental organizations, making its evaluations available to a proposed central office in the State Department.

UNITED NATIONS OCEANOGRAPHY (p. 23)

The ocean science programs of the UN system should be combined in a single separate organization to deal with scientific and engineering aspects of intergovernmental ocean affairs. This organization should provide technical support to organizations of the UN system and other bodies concerned with the political aspects of ocean affairs and with development and management of ocean resources and should coordinate relevant programs with those of non-UN organizations.

INTERNATIONAL SCIENTIFIC ORGANIZATION (p. 25)

The interrelation of ocean activities among the various international nongovernmental scientific bodies should be affected by a strengthened SCOR. In the United States, the National Academy of Sciences Ocean Affairs Board should be the focal point for relations with these international organizations.

GLOBAL MARINE SCIENCES SERVICES (p. 33)

With the assistance of the National Academy of Sciences and the National Academy of Engineering, NOAA should review the various national and international proposals for global monitoring and environmental forecasting in the atmosphere and ocean (including the biosphere) and for evaluating the effects of pollutants and should assess the feasibility of developing a comprehensive and unified system for these purposes. The National Oceanic and Atmospheric Administration should also assume the principal U.S. responsibility for promoting international data exchange and for strengthening charting and navigational aids for international use.

STATE DEPARTMENT CENTRALIZATION (p. 55)

Within the State Department, a single office should be assigned responsibility for developing a coherent ocean policy, for maintaining U.S.

positions consistent with that policy in the various intergovernmental bodies, and for coordinating and supervising pertinent activities of offices within the Department concerned with ocean affairs. This office should also compile, analyze, and promulgate information required for rational decision-making on ocean problems.

FEDERAL REORGANIZATION (p. 56)

For the purpose of coordinating activities of NOAA with the ocean programs of the Department of Health, Education, and Welfare, the Atomic Energy Commission, the National Science Foundation, the Department of Defense, and other federal agencies, an interagency committee on ocean affairs should be established under the chairmanship of the Administrator of NOAA. The regrouping of federal ocean-oriented agencies in NOAA should be reinforced by rationalization of relevant committee structure in Congress.

NATIONAL FISHERY REGULATION (p. 56)

All marine fisheries subject to U.S. jurisdiction should be regulated by an appropriate federal agency, at present the National Marine Fisheries Service.

APPRAISAL OF DECISIONS (p. 57)

The systematic appraisal of the effect of decisions on the achievement of ocean policy objectives should be carried out by an appropriate federal body.

INTERNATIONAL FISHERY REGULATION (p. 59)

A comprehensive study of the nature and effectiveness of actual and possible arrangements for international fishery regulation should be made by a nongovernmental group. The study should include such problems as methods for obtaining the necessary scientific data, scope of power vested in international fishery bodies, breadth of membership of these bodies, the utility of bilateral agreements, and the potential usefulness of a global regulatory agency. The results of this study should be used by the State Department as a basis for U.S. ocean policy.

FREEDOM OF SCIENTIFIC RESEARCH (p. 72)

The United States should act to secure the maximum freedom of access for scientific exploration and research in all parts of the world ocean. A variety of international mechanisms for removing restrictions on research should be used as necessary. Revisions of the Continental Shelf Convention to increase restrictions on research should be resisted; preferably, revisions should eliminate the present "consent" requirement.

Any regime for resources of the seabed beyond the limits of national jurisdiction should involve minimal interference with scientific research. The United States should oppose allocation of exclusive rights of exploration of this seabed.

Freedom of scientific exploration and research should be considered an integral part of the doctrine of freedom of the seas. Scientists should oppose restraints on marine science that derive from agreements on other uses of the sea.

OCEAN DATA ACQUISITIONS SYSTEMS (ODAS) TREATY (p. 76)

The United States should actively support a treaty to protect unmanned buoys, free-floating instrument packages, and other ocean data acquisition systems (such as ODAS) now in use and under development. To expedite agreement on an ODAS treaty, the United States should resist all attempts to tie this question to broader problems of the law of the sea, including the limits of national jurisdiction.

UNILATERAL ACTION (p. 76)

The United States should make a unilateral declaration allowing scientific research in areas outside internal waters but subject to its jurisdiction, provided certain conditions are observed.

TECHNICAL ASSISTANCE (p. 86)

An appropriate component of the National Academy of Sciences should undertake a comprehensive study of the nature, scope, and effectiveness of various attempts to provide technical assistance to developing countries in marine science and resource development and in other aspects of ocean affairs.

CHAPTER 2

Ocean Research

Significant gains in understanding in two general areas of ocean research can be achieved through positive and active international cooperation. Investigations of processes that are of ocean dimensions can be advanced through such cooperation because of the vast scale involved and the relative slowness with which observations can be made. In applications-oriented research, international cooperation is essential to establish an agreed basis for rational use of the ocean and its resources both nationally and internationally.

Processes of ocean dimensions are those with a scale comparable to that of major portions of the world ocean, measured in thousands of kilometers. In time they may extend from months and seasons through decades to the countless years of geological change. Examples of these large processes are common in studies of the ocean floor, e.g., sea-floor spreading, sediment transport and deposition, subsidence, uplift, plate movement, and other tectonic phenomena. They are also common in studies of the major ocean circulation gyres and in the interactions with the atmosphere that affect seasonal and climatological phenomena. These processes control the distributions of chemical species (in-

9

cluding pollutants) and the distributions of productive regions and of living organisms throughout the world ocean.

Some aspects of these problems, such as those of a theoretical nature, are best studied by individual scientists and institutions. The principal need for international cooperation, then, is in the efficient exchange of data, information, and ideas and in mutual planning for the acquisition of new information.

Often, however, further knowledge of large oceanic processes depends critically on broad-scale observation programs or surveys. Effective measurement may require that the process or phenomenon be observed in many places (to represent the spatial variations) within a time that is short relative to the temporal variations. If the time variations are to be accurately portrayed, it may be necessary to repeat such synoptic observations frequently over a long period of time. Such a matrix of observations may require an effort that is greater than one nation is able to assume. In ocean floor studies, although time changes are slow, large cooperative efforts may still be required to cover the vast areas in sufficient detail and in finite time.

These activities are of interest not only to the ocean scientist but also to those who seek benefit from ocean use. Activities involving the use of the ocean and its resources include the extraction of living and mineral resources; shipping; navigation; coastal works; siting and maintenance of cables, pipelines, and tunnels; disposal of wastes; forecasting oceanic and atmospheric conditions; extraction of energy; military operations of many kinds; and recreation.

To some extent, all ocean research can be related to one or another of these applications. Certain investigations that are directly related to use of the ocean and its resources depend heavily on international action for their success. These include

- Exploration and assessment of resources
- Study of processes affecting the abundance, distribution, and availability of resources
- Studies required for resource management and conservation
- Studies of the interactions between uses of resources, and other effects of human intervention
- Studies leading to useful predictions of oceanic and atmospheric conditions

The nonliving resources of the continental slopes and the deep ocean floor are poorly known. The world community requires more information to make proper decisions on policies, principles, and procedures for rational use of such resources. The exploration and assessment of these resources can most effectively be accomplished through international cooperation, both in surveys at sea and in compilation and evaluation of available data.

Living resources differ in that their abundance, distribution, and availability are highly variable. Many species are distributed across and beyond boundaries of national jurisdiction and are harvested by citizens of several nations. Exploration and assessment of specific resources can be accomplished by well-designed cooperative studies.

Understanding of processes affecting the abundance, distribution, and availability of living resources is vital to the management and conservation of these resources. The studies involve measures of the exploited population and the effort applied in its exploitation as well as the environmental changes to which it is exposed. *International cooperation is required, both in the compilation and exchange of information and in the organization of field investigations.*

Resource management decisions are not based solely on scientific considerations and cannot be. There is a parallel set of investigations required in such fields as economics, sociology, political science, and law, if resources are located beyond the limits of jurisdiction of a single nation, if they are affected by events occurring beyond those limits, or if their use affects more than one country. International cooperation in social studies on appropriate aspects of the use of the ocean and its resources is needed and should be encouraged.

Not all uses of the ocean are compatible, and increasing oceanic activity leads to the increased possibility of conflict. For example, the extraction of mineral resources may interfere with fishing, or vice versa. Man's use of the ocean as a receptacle for waste products may have profound effects on the biosphere or may damage the recreational potential of the ocean.

Several kinds of oceanographic investigations are required if man is to control modification of the ocean environment or to predict the consequences of interactions among different ocean uses. The processes whereby introduced substances are distributed, transformed, degraded, or concentrated in the ocean need to be studied. Base lines against which the effect of interventions can be measured must be established, and substances introduced into the ocean through the action of man must be monitored.

All of these studies will benefit from international cooperation in formulation of the investigations, and the international community will benefit from the exchange of information on their findings. An adequate monitoring program will require closely coordinated joint action on the part of all maritime nations.

Most uses of the ocean and its resources are subject to the vagaries of ocean and atmosphere. Although the control or deliberate modification of oceanic atmospheric conditions may be far in the future, even the successful prediction of events could bring immense economic benefits. The development of systems for monitoring and prediction depends on a wide variety of investigations.

As is true of most ocean research, these investigations must be carried out at all levels of intensity, from the individual scientist or laboratory to the multinational experiment. Ultimately, observing networks or systems will be deployed over large regions or on an oceanwide or global basis. Thus, international action is inherent in the development of these large-scale networks. During the development phase, critical field experiments and installation of pilot arrays particularly will benefit from international cooperation. Operation of regional or ocean-wide systems will normally be an international function.

MECHANISMS FOR INTERNATIONAL ACTION

International action is effected by both national and international institutions. Mechanisms are required for funding, for planning and coordination, and for communication. Organizations that serve these purposes fall into two general categories, nongovernmental and governmental (public). In most cases, funding is provided by governmental organizations, planning and coordination by both governmental and nongovernmental organizations, and communications predominantly by nongovernmental organizations.

FUNDING MECHANISMS

In the United States, as in most other nations, most oceanographic research is financed by the government. Although there are also state-level public agencies concerned with some aspects of marine research, these are generally not a major element in international programs; they can, however, make an important contribution to resource-relevant investigations. A limited amount of private funds also supports academic research. Private ocean research is also conducted by industry, most notably by the petroleum industry. However, the results of such research do not often become available to the international community except when commercial exploitation is a consequence of the re-research. Thus, from the international point of view, private funding can be largely ignored.

Within the federal government, a number of operating agencies concerned with some aspect of the marine sciences, including the former Environmental Science Services Administration (ESSA) and the Bureau of Commercial Fisheries (BCF), have recently been brought together in NOAA. Large in-house programs are also maintained by the Navy. Most research in academic laboratories is funded by the National Science Foundation and the Office of Naval Research; NOAA is involved in academic research through its Sea Grant Program. Other important participants in the federal oceanographic program include the Coast Guard, Atomic Energy Commission, Smithsonian Institution, Public Health Service, and Geological Survey.

On the international level, funding mechanisms serve a different purpose, since the principal expenses are met nationally. In an international cooperative expedition, for example, the principal international expenses are consultation and coordination and the publication of results.

A few intergovernmental organizations conduct marine investigations directly with their own personnel, including the Inter-American Tropical Tuna Commission, the International Pacific Halibut Commission, the International Pacific Salmon Commission, and the International Atomic Energy Agency (through its Monaco laboratory). But, generally, scientific programs in which intergovernmental organizations are concerned are carried out by their member states rather than by the organization itself, the latter serving primarily as a coordinating mechanism. As funding mechanisms, intergovernmental organizations are much more concerned with providing technical assistance to developing countries, largely in connection with the development of marine resources, than with the direct support of marine research.

Some general comments on the relative roles of developed and developing countries in international cooperative investigations should be made here. Only a few industrialized countries have a major capability in large-scale modern oceanographic research. Only these countries have a sufficient number of trained marine scientists and funds and facilities for carrying out broad work in marine science. Of these countries, an even smaller number has enough oceanographic strength to engage in research in distant waters. The smaller industralized countries devote their efforts principally to work in the vicinity of their coasts or in nearby seas.

The developing countries have limited facilities, restricted funds, and few trained scientists. The limited resources available are usually devoted to the study of local problems. Thus, the contribution of developing countries to international cooperative investigations is often confined to joint studies with visiting investigators and to operation of monitoring programs that can provide knowledge of local spatial variations to supplement broader observations.

A substantial part of the budget and staff of the major specialized United Nations agencies, such as the United Nations Education, Scientific and Cultural Organization and the Food and Agriculture Organization, is applied to problems of developing countries. Much larger sums are available for resources development from the United Nations Development Program for which the specialized bodies of the United Nations serve as executing agencies. Only a small part of the UNDP programs in the marine area is devoted to scientific investigation of the ocean, the major part being allocated to technological, economic, and training aspects.

An important contribution of UNESCO to science in general, as well as to marine science, should be recognized. For some years, UNESCO has been a major source of income to the International Council of Scientific Unions and its constituent bodies, including the Scientific Committee on Oceanic Research. Although the sums are not large relative to the cost of doing research, they have been an invaluable help in the functions of scientific communication and, to some extent, in planning and coordination.

In comparison with the large UN organizations, intergovernmental bodies such as the Intergovernmental Oceanographic Commission, the International Council for the Exploration of the Sea, or the International Hydrographic Organization have extremely small budgets and staffs. Most work is carried out and funded by member states in accordance with agreements reached at regular legislative assemblies.

PLANNING AND COORDINATION MECHANISMS

Planning of oceanographic research programs in the United States is highly decentralized, involving many separate individuals, laboratories, and agencies.

The planning and coordinating functions are carried out by organizations that are often referred to as councils. In the United States, the Executive Office assigned the major role in coordinating the marine programs of various federal agencies to the National Council of Marine Resources and Engineering Development.* This coordination is difficult because of the large number and diversity of agencies concerned and because funding levels of the individual agencies are established within their parent departments, in the Office of Management and Budget, and through a variety of committees in the Congress. In practice, the compelling priorities of the constituent agencies prevent the development of a truly integrated national program. To many, it has appeared that effective coordination could result only when amalgamation of the major participants takes place. It remains to be seen if the formation of NOAA will lead to a significant improvement.

There is no overall coordination within the U.S. Government of the international components of federal oceanographic programs. Marine research relating to international fisheries has been coordinated through the Bureau of Commercial Fisheries (now the National Marine Fisheries Service) with generalized supervision being exercised through the office in the Department of State that bears responsibility for negotiating international arrangement for fisheries management. The Special Assistant to the Secretary for Fisheries and Wildlife coordinates the research programs that bear on fisheries management in areas of interest to the United States with the responsible office in NMFS.

Some aspects of other international components of federal programs have been coordinated by the State Department Committee on International Ocean Affairs, although virtually all this activity is carried out by a working-level group, known as the Panel on International Programs and International Cooperation in Ocean Affairs PIPICO. Representatives of the various federal operating agencies participate in these activities to advise the State Department about relevant agency programs and to coordinate them to the extent practicable. As might be expected, purely technical aspects are coordinated with less difficulty than those aspects, including technical ones, that have political implications. Difficulties arise because of the number of offices in the State Department concerned with various aspects or ocean affairs, none of which has the responsibility for developing

---

*After completion of this report, the Council was terminated. The coordinating function performed by the Council is being carried out by the Interagency Committee on Marine Science and Engineering, reporting to the Federal Council for Science and Technology, and by the National Advisory Committee on Oceans and Atmosphere, reporting through the Secretary of Commerce to the President and the Congress.

a coherent U.S. ocean foreign policy. One or another office domi-
nates, depending on whether the UN General Assembly, the Intergovernmental Oceanographic Commission, one of the specialized agencies
of the UN, a fishery commission, the North Atlantic Treaty Organization, or some intergovernmental body is involved, leading to the multi-
faceted and ever-changing appearance of present policy. This problem
is examined further in the section discussing the mechanisms for prescribing for rational use of the ocean.*

Comparable intergovernmental councils have the function of
coordinating national activities or those of other intergovernmental organizations. Because ocean research is conducted primarily at the national level, the former task has the more important consequences for oceanography. The oldest of such councils
is the International Council for the Exploration of the Sea, which for
nearly seventy years has provided a mechanism for joint action among
European countries working in the North Atlantic. Although regional
in character and principally concerned with fishery-related problems,
ICES has had remarkable success in planning and coordinating scientific investigations. The effectiveness of ICES is often cited as proof of
the desirability of regional, as opposed to global, action, but it should
be noted that membership of the organization consists almost exclu-
sively of highly developed countries with great competence in oceanography and long experience in cooperative activity in many fields.
Other regional bodies have had less success in promoting scientific investigations, largely because they have not commanded the scientific
resources necessary for the task. However, regional fishery commissions in the Northwest Atlantic and North Pacific have been especially active and effective in coordinating significant scientific programs
concerned with living resources.

By the early 1950's, the earth sciences had advanced to the point
where it seemed desirable to develop a global program to apply the
newly available methods and concepts to the study of the planet as a
whole. The resulting International Geophysical Year was organized
and coordinated by the nongovernmental International Council of
Scientific Unions. Although IGY was a success, it became evident that
personal arrangements among scientists were not a sufficient basis for
such immense campaigns and that mechanisms were required for inter-

*Since this report was written, the Department of State has partially recognized the problem
and has acted upon it by naming a Coordinator of Ocean Affairs to act as deputy to the
Undersecretary for Political Affairs. It remains to be seen whether this move will in fact
provide the basis for developing and maintaining a truly coherent ocean policy.

governmental agreement and support for scientific programs. The International Indian Ocean Expedition, initiated by the Scientific Committee on Oceanic Research late in the decade, further illustrated the practical difficulties in organizing large-scale projects without formal governmental commitments.

Through the initiative of UNESCO, a series of conferences led to creation in 1960 of the Intergovernmental Oceanographic Commission, a council designed to "promote scientific investigation with a view to learning more about the nature and resources of the oceans through the concerted action of its Members."[1] Since then, IOC has demonstrated the utility of such a mechanism in its organization and coordination of a number of cooperative investigations (including the International Indian Ocean Expedition in its later phase, the International Cooperative Investigation of the Tropical Atlantic, the Cooperative Study of the Kuroshio and Adjacent Regions, and the Cooperative Investigation of the Caribbean and Adjacent Regions), in its promotion of intergovernmental agreements on the marking of oceanographic buoys and on the allocation of radio frequencies for oceanographic purposes, in its studies of legal problems related to ocean data systems and to freedom of scientific research and in its planning for the Integrated Global Ocean Station System.

There are also more specialized intergovernmental councils, such as the International Hydrographic Organization, which coordinates the international aspects of sea floor charting, including some oceanographic work of national hydrographic organizations. There are numerous intergovernmental fishery commissions concerned with conservation of fishery resources in various regions of the ocean. Some of these commissions, such as the International Commission on Northwest Atlantic Fisheries, have organized international cooperative investigations similar in scale to those of IOC or ICES.

As noted earlier, intergovernmental councils also exist for coordinating the work of other intergovernmental organizations. Until recently, there has been only one such council in the field of ocean affairs, the Sub-Committee on Marine Science and Its Applications of the Administrative Committee on Coordination, Economic and Social Council of the UN. The ACC, composed of the administrative heads of the several UN bodies, has the function of coordinating activities within the UN system, and the Sub-Committee has provided the means for consultation at the working level among representatives of the various agencies. It has not been a very effective council. Agency representation is below the policy-making level, and, because of different budgetary periods and practices among the agencies, coordination has been limited to the exchange of information on plans and programs.

New and potentially more effective mechanisms are now being developed through the IOC. These result from the action of the UN General Assembly in approving a long-term and expanded program of oceanographic research (the so-called Expanded Program, or LEPOR) and in asking IOC to intensify its activities with regard to coordinating the scientific aspects of the program. Pressure has also developed from other UN bodies in addition to UNESCO for a stronger role in supporting IOC. This has led to establishment of an Intersecretariat Committee on Scientific Programs Related to Oceanography. The evolution of IOC now under way may lead to more effective means of coordinating the scientific ocean activities of the interested UN organizations.

Mechanisms for coordinating the activities of one or another UN body with those of intergovernmental organizations outside the UN system are being developed on an experimental and *ad hoc* basis. For example, IOC and FAO are cooperating with ICES in the development of a Cooperative Investigation of the Northern Part of the Eastern Central Atlantic and with the International Commission for the Scientific Exploration of the Mediterranean in the development of the Cooperative Investigation of the Mediterranean. However, no permanent mechanisms for this kind of coordination have been established. Because a given country may belong to a number of organizations, some coordination might be expected to take place at the national level (i.e., each country would implement different aspects of its single national ocean policy through the different organizations). This does not often happen, perhaps because national links with the various organizations are weak. The problem will undoubtedly become more critical with development of the Expanded Program (LEPOR), which should encompass international cooperative programs of many sorts, not just those developed within the UN system.

Nongovernmental councils exist at both the national and international levels. In the United States, the principal council of this sort is the National Academy of Sciences's Ocean Affairs Board (formerly the Committee on Oceanography). Other such councils, with more specialized interests, include the Committee of Oceanographic Laboratory Directors and its subsidary, the Research Vessel Operators Council.

An important role of the Ocean Affairs Board (formerly the Committee on Oceanography) is that of providing advice to government agencies and councils. Other tasks include the evaluation of progress in the ocean sciences and serving as the U.S. National Committee to SCOR. With the assistance of its Committee on International Marine Science Affairs Policy, the Board helps to determine U.S. participation and positions in certain intergovernmental bodies through its advisory relationship with the Department of State, thus providing a means whereby scientists can influence the U.S. role in international ocean affairs.

These nongovernmental councils contribute to the planning and coordination function largely through their advisory actions. This activity calls for some general comments. One can distinguish between two broad types of advisory bodies. An advisory body of the first type is independent of the body being advised, has other responsibilites in addition to providing advice, and should be able to provide general advice on scientific questions as well as to monitor and comment on scientific activities of the body being advised. An advisory body of the second type is dependent, has the unique responsibility of providing advice on questions raised by the body being advised, and should be able to provide detailed suggestions on the programs of the parent body. This type of advisory group is the more common of the two and calls for no particular discussion here.

Advisory bodies of the first type, such as the Ocean Affairs Board, provide an essential link between individual scientists and public programs. Decisions on public policy must usually be based on more than technical aspects. Policy makers must be fully informed on the technical aspects and the consequences of alternative choices. This information has the greatest chance of being objective if it is provided by independent experts. Such advice is useful (even when unsolicited) and should arise, in part, from a continual monitoring of public programs. Even when the advice is directed to national programs, it may have international consequences to the extent that national programs are integrated in international cooperative investigations. From this point of view, an important function of the nongovernmental council is to make known the views of scientists to those charged with developing national ocean policy.

International nongovernmental councils may contribute to the planning and coordination of intergovernmental programs by monitoring and evaluating such programs and by presenting their views to the responsible bodies. In addition, such councils may organize limited cooperative programs of their own. Such programs may consist, for example, of field trials for the intercomparison of equipment or methods or joint studies by several laboratories of a specific region or problem. Large-scale investigations, as noted earlier, appear to be more effectively arranged through intergovernmental agreements.

There are several councils with ocean interests within the International Council of Scientific Unions. The principal ocean-oriented council is the Scientific Committee on Oceanic Research, which is supported by contributions from its affiliated national committees and by UNESCO through an annual contract. The dual nature of SCOR's activities reflects these two sources of support. In part, SCOR responds to the initiatives of its members and national committees in establishing working groups or other activities for the improvement or intercomparison of methods or for the consideration of selected scientific problems. As a scientific body advisory to UNESCO and IOC, SCOR responds to initiatives of these organizations by establishing appropriate subsidiary bodies, by correspondence with national committees, and through consideration by members of the SCOR Executive Committee.

Within ICSU there are other interdisciplinary bodies with ocean interests that function to some extent as councils. The Scientific Committee on Antarctic Research has a subsidiary group concerned with Antarctic oceanography. The Special Committee on the International Biological Program has a section of marine productivity, and a Special Committee on Problems of the Environment has been established. There are also interunion bodies, such as the Upper Mantle Committee or the Commission on Geodynamics that may develop scientific programs with a significant ocean component. In general, these bodies work closely with SCOR on oceanographic matters.

In its relation with UNESCO and IOC, SCOR plays the part of an advisory body of the first class, as described earlier. A similar function is performed by the FAO Advisory Committee on Marine Resources Research when it serves as an advisory body to IOC on the fishery aspects of its programs. When advising FAO, on the other hand, ACMRR is an advisory body of the second class. A similar dual role seems to be intended for the new WMO Advisory Committee on Oceanic Meteorological Research. IOC is also considering appointment of an advisory group in ocean engineering like the recently established Engineering Committee on Oceanic Research.

COMMUNICATION MECHANISMS

Scientific organizations were first created to improve communication among scientists, through the convening of scientific meetings and the publication of scientific papers. In most countries where ocean research is conducted, there exist such professional and technical societies, either specifically intended for oceanographers or, more commonly, as sections of organizations concerned with more basic disciplines. In the countries most active in ocean research, there are innumerable opportunities to discuss and publish scientific results. In the United States, societies more directly concerned with marine science include the American Society of Limnology and Oceanography and the Oceanographic Section of the American Geophysical Union. Other societies include the Marine Technology Society and the American Fisheries Society, both concerned more with applications than with science *per se.* A number of other societies with broader interests often schedule scientific discussions on ocean topics.

Internationally, ocean-oriented societies are, for the most part, components of the International Council of Scientific Unions. The principal ones are the International Association for the Physical Sciences of the Ocean of the International Union of Geodesy and Geophysics, the International Association of Biological Oceanographers of the International Union of Biological Sciences, and the Commission on Marine Geology of the International Union of Geological Sciences. Other societies with different or broader interests often consider oceanographic topics. Discussions on interdisciplinary topics are also organized by nongovernmental councils such as SCOR and by agencies like UNESCO and FAO. Two international oceanographic congresses have been held and the Joint Oceanographic Assembly was arranged in 1970 by SCOR and a number of ICSU societies.

Communication does not seem to be inhibited by the inability of scientific societies to organize meetings. It is even conceivable that too many meetings are being placed on the scientists' agenda. But the important international scientific meetings are often seriously underfunded, and it is usually difficult for scientists to finance attendance at distant meetings.

RECOMMENDATIONS CONCERNING U.S. POLICY

In general, the national and international mechanisms referred to previously affect not only ocean research but also other aspects of ocean affairs. The basic argument for change in some of these mechanisms is made elsewhere in this report, and comments here are restricted to the consequences for ocean research. In other cases, the principal arguments are presented in this section.

INTERNATIONAL ROLE FOR NOAA

The simplification of the federal oceanographic structure resulting from the creation of NOAA should greatly strengthen the capability of the United States to participate in international cooperative investigations. The Navy has been the largest single agency engaged in oceanographic work, but for political reasons it is usually difficult to have the Navy represent U.S. interests in international civilian programs.

NOAA should serve as the focal point for U.S. participation in international ocean investigations. By bringing together the interests of a number of formerly individual agencies, such as BCF and ESSA, it should be possible to prepare a stronger case for this participation. Centralization of agency responsibility for U.S. participation should vastly improve coordination both among participating groups in this country and between the collective effort of this country and that of other participating nations.

Because of its competence in a variety of scientific and applied fields, the NOAA staff should be well qualified to keep track of the technical aspects of problems considered by the various ocean-oriented intergovernmental organizations. Although the State Department has the responsibility for relations with these organizations, it is impracticable for it to develop the same technical competence as that of NOAA. Yet many political decisions should be based on understanding of the technical aspects.

An office within NOAA should keep under review all documents and reports on relevant scientific and engineering matters dealt with by IOC, WMO, FAO, IMCO and other intergovernmental organizations, making its evaluations available to a proposed central office in the State Department.

STATE DEPARTMENT CENTRALIZATION

The absence of a U.S. ocean policy and the diffusion of responsibilities for ocean affairs within the State Department works to the detriment of an effective U.S. role in scientific investigation of the ocean. Many of the problems could be solved with establishment of a central office of ocean affairs within the Department of State, as proposed elsewhere in this report. An interagency group such as PIPICO would still be necessary for developing the details of U.S. positions, and means should be developed for a more effective input from marine scientists to this group.

UNITED NATIONS OCEANOGRAPHY

Although several of the UN organizations support or coordinate ocean-ographic programs, none is principally concerned with science, let alone ocean science. Funding of these programs has remained essentially level, and there is little opportunity for significant growth within the present organizational structures. More serious is the fact that, because the agencies have focused their attention on other matters, their support for new developments and opportunities is difficult to obtain.

Even the present funds could be used more effectively if the programs were combined rather than being distributed among several agencies. This would permit elimination of overlap in staff activities, mutual reinforcement of strengths in staff and programs, and combining of certain program elements such as fellowships, training activities, and publications.

Ultimately, a separate intergovernmental agency concerned with the scientific and engineering aspects of ocean affairs, including ocean resources, is needed. This might be a specialized agency of the United Nations that would, among other things, provide technical support to the United Nations or to any other body concerned with the political aspects of ocean affairs and with the development and management of ocean resources. Initially, it might include the IOC, the UNESCO program in oceanography, the World Data Center System for oceanography, the FAO Department of Fisheries, and ocean elements of the Division of Natural Resources and Transport of the UN secretariat. The agency would require its own competence in meteorological oceanography and should work closely with WMO on problems of mutual interest. The addition of other activities by such an agency should be considered. For example, the International Hydrographic Organization and its responsibilities for coordination of charting and other navigational services could be included in the agency.

The ocean science programs of the United Nations system should be combined in a single separate organization to deal with scientific and engineering aspects of intergovernmental ocean affairs. This organization should provide technical support to organizations of the UN system and to other bodies concerned with the political aspects of

ocean affairs and with development and management of ocean re-
sources, and should coordinate relevant programs with those of non-
UN organizations.

Such a new agency should provide an eventual opportunity for a
new level of funding. In the absence of a suitable political climate for
establishing the agency, an immediate expedient would be to draw
FAO and WMO more directly into participation in, and support of,
IOC work. In this way, IOC would form the seed crystal for later development of the new agency. Steps in this direction are already
being taken.

Whether UN oceanographic programs are organized through the
IOC or a new agency, there will still be a need to coordinate with
programs of organizations outside the UN system. Countries that are
members of IOC may still wish to carry out certain investigations
through ICES or ICNAF, or they may make special bilateral arrangements for a specific investigation (as was done with the EASTROPAC*
Expedition). The results of effort expended in such programs are of
as much interest and value to IOC as those of its own programs.

Until recently, there has been little experience in this type of coordination. As noted earlier, several *ad hoc* arrangements are being
tried on an experimental basis, and such arrangements should be
evaluated before a decision is made on the preferred coordinating
machinery. It may eventually be desirable for IOC, or a new agency
that replaces it, to establish more formal links with these non-UN
bodies, to enhance consultation and joint action with non-UN bodies,
on projects of mutual interest.

INTERNATIONAL SCIENTIFIC ORGANIZATIONS

As discussed before, there are several different kinds of organizations
within the International Council of Scientific Unions that deal with
one or another aspect of marine science. These range from unions concerned with the basic disciplines of science, through specialized societies to interdisciplinary committees and commissions. Their functions
include planning, coordination, and communication.

Because of the voluntary character of ICSU and its constituent
bodies, no division of responsibilities can be absolute. The ICSU body
most clearly oriented toward the ocean is the Scientific Committee on
Oceanic Research, made up of representatives of ocean-oriented coun-

---

*An oceanographic study of the Eastern Tropical Pacific.

tries as well as marine-centered components of ICSU bodies such as the International Association for the Physical Sciences of the Ocean of the IUGG, the Section on Oceanography of the IUBS, and the Commission on Marine Geology of the IUGS. SCOR recently revised its constitution to permit a broader and more effective coordinating role.

The interrelation of ocean activities among various international nongovernmental scientific bodies should be effected by a strengthened SCOR. The national links with international nongovernmental bodies in the ICSU family are centered in the National Academy of Sciences and the Divisions of the National Research Council. In the case of ocean-oriented bodies, these links are diffuse and inadequately co-ordinated. For example, the Ocean Affairs Board serves as the National Committee to SCOR, while the oceanographic section of the American Geophysical Union is the corresponding body to IAPSO (as is the AGU to the parent union, IUGG). With the exception of the OAB–SCOR relationship, these links are monodisciplinary in nature, as is the basic structure of ICSU itself.

In the United States, the NAS Ocean Affairs Board should be the focal point for relations with the international scientific organizations interrelated through SCOR. When the Board meets as National Committee to SCOR, representatives of U.S. corresponding bodies to relevant international organizations should be invited to participate.

CHAPTER **3**

Provision of
Global Marine
Science Services

The report of the Stratton Commission, *Our Nation and the Sea,*[2] noted that a broad spectrum of activities in the U.S. ocean program depends on technical support, including not only data management but also instrumentation services, mapping and charting, aids to navigation, accurate forecasts, and search and rescue programs.

With the increase in global ocean activities, there is a growing need for global marine science services. Such services are of particular importance to the success of the international marine research program of the Expanded Program (LEPOR), including the International Decade of Ocean Exploration.

A recent and valuable definition of activities involving global services of value to international marine research is in the report, *International Ocean Affairs,* produced in 1967 by a nongovernmental joint working group of the Scientific Committee on Oceanographic Research, the Advisory Committee on Marine Resources Research, and the World Meteorological Organization.[3] This report identifies the following services as valuable:

A. *Exchange of Information*
   1. Transmission, storage, exchange, retrieval and processing of data

2. Storage, exchange, retrieval, indexing, abstracting and translating of literature and other documentation

3. Exchange, sorting and storage of specimens, samples and other materials . . .

6. Exchange of information about research programs, scientists, institutions and facilities

B. *Standardization and Intercalibration of Methods*
   1. Method development and testing
   2. Reference methods
   3. Intercalibration experiments
   4. Provision of standard equipment
   5. Standards laboratories, national and international

C. *Cooperative Investigations and Assessments*
   1. Characteristics normally independent of time (including charting of the sea floor and surveys of non-living resources)
   2. Time-dependent characteristics, including scientific bases for ocean forecasts
   3. Regional and worldwide networks of ocean data stations and observations . . .

F. *Technical Services*
   1. Navigational aids for research and marine activities
   2. Assignment of radio frequencies
   3. Forecasting and warning services—such as tsunami, hurricane or storm surge warnings; forecasts of sea surface temperature, sea state or ice conditions.[3]

The IOC Comprehensive Outline of the Expanded Program includes data on information management, instrumentation and methods, and "the widespread availability of precise navigational systems, improved communications, more complete and accurate forecasts of the marine environment, and the expansion of programmes of hydrographic surveys, mapping and charting."[4] It states that the implementation of the Integrated Global Ocean Station System will draw heavily on all the supporting activities of the Expanded Program (LEPOR), with particular emphasis on

Development of appropriate technology and instrumentation, standardization and unification of instruments and methods of observations for the IGOSS programme;
standardization of procedures for use of the radio-telecommunications channels;
organization of the oceanographic service in an integrated fashion and patterned after the World Weather Watch[5]

There are several areas in which considerable progress has already been made toward the provision of global services. Probably the area of most extensive coverage is marine meteorology, with provision of weather information through the facilities of the World Meteorological Organization. This is a truly international and truly global service. The data provided do include some ocean information, such as sea surface temperature, and the WMO apparently is anxious to cooperate in the establishment and improvement of global services for marine activities.

A second area in which a program of global services has made a small start is marine fisheries. World catch statistics, as well as a few other services, are provided by the Food and Agriculture Organization of the United Nations.

Beginning with the International Geophysical Year in the late 1950's, World Data Centers were established for the provision of global services. The report, *Guide to International Data Exchange,* prepared by the International Council of Scientific Unions in 1963 states:

This new form of international cooperation—exchange of data through WDC's—was found to be very effective. Instead of having to address themselves to many national organizations, scientists could receive data necessary for scientific work directly from the WDC's. Thanks to the collection and exchange of data through WDC's, it became possible to investigate phenomena on a planetary scale and to study the interdisciplinary relationships among various phenomena.

Each WDC is responsible for: (1) endeavoring to collect a complete set of data in the field or discipline for which it is responsible, (2) the safekeeping of the incoming data, (3) correct copying and reproduction of data, maintaining adequate standards of clarity and durability, (4) supplying copies to other WDC's of data not received direct, (5) preparation of catalogues of all data in its charge, (6) making data in the WDC's available to the scientific community. *In particular,* WDC's are required to supply copies of material in the data center to any scientific body or investigator in any country (for a cost not to exceed the cost of copying and postage) and, by appropriate arrangement, to enable scientists to work directly with the materials in the WDC's.

The World Data Centers collect data and publications in the following disciplines: . . .

II. Meteorology, synoptic and physical (atmosphere density, atmospheric radioactivity, radiation, ozone, noctilucent clouds, atmospheric electricity, thunderstorms, . . .
    III. Geomagnetism, earth currents, and paleomagnetism
    IV. Aurora and airglow
    V. Ionosphere
    VI. Solar activity
    VII. Cosmic rays
    VIII. Longitude and latitude
    IX. Glaciology

X. *Oceanography* (italics added)
XI. Rockets and satellites
XII. Seismology, including tsunamis
XIII. Gravimetry, earth tides, recent movements of the earth's crust
XIV. Physics and chemistry of the earth's interior (volcanology, geothermics, properties of rocks under conditions of high temperatures and pressures, absolute age determinations of the earth)[6]

In the United States, World Data Center A for Oceanography is collocated with the National Oceanographic Data Center, and together these organizations receive, compile, process, and preserve oceanographic data from both foreign and domestic sources and disseminate these data to U.S. and foreign marine science organizations.

The International Hydrographic Organization (then called the International Hydrographic Bureau) was founded in 1921 and provides international hydrographic services to

Establish close association between national hydrographic offices;

Encourage adoption of the best methods for carrying out hydrographic surveys and coordinating hydrographic undertakings with a view to rendering navigation easier and safer; and

Obtain uniformity in hydrographic documents for efficient international use.

Presently, the Intergovernmental Oceanographic Commission and the World Meteorological Organization are jointly planning the Integrated Global Ocean Station System. When it is in operation, IGOSS will constitute an important global service. Its purpose is the rapid collection and dissemination, on a worldwide basis, of ocean data that must be exchanged in real time in order to describe the present conditions of the ocean and enable forecasters to predict changes. As described in the Report of the Panel on Environmental Monitoring of the Stratton Commission,

Tentative plans call for the IGOSS to include the following components:

- An observational network comprising all types of ocean data stations and observational techniques:
  - automatic telemetering buoys
  - coastal stations and research vessels
  - fixed off-shore platforms
  - observational satellites
  - other new means that may be developed.

- A communication service for data transmission.
- Centers for collection, processing, retrieval, and dissemination of data.[7]

A subsequent analytical description was offered in the joint IOC/WMO paper of June 1969:

While oceanography and meteorology have already progressed a long way towards understanding the oceanic and atmospheric processes, a more systematic, multidisciplinary approach is needed to provide the data necessary for continuous monitoring of oceanic conditions on a global scale and for improving and/or developing reliable methods of forecasting.

What has been lacking so far is an approach whereby the environmental characteristics of both the ocean and the atmosphere could be observed and measured at a sufficient number of points to permit a coherent analysis of the processes therein and of their interactions, both at any given moment and in their evolution. . . .

Member States have expressed interest in IGOSS as a major contribution to a support service for navigation and maritime commerce, fishing and petroleum industries, coastal warnings, health and recreational activities, and various research projects. Improved oceanographic and meteorological operational data are expected to increase considerably the accuracy of long- and short-term weather and oceanographic forecasts; observations from fixed points will serve as reference stations for assembling and interpreting data from meteorological satellites. The data from global observations will provide a rich source for the oceanographic and meteorological sciences, allowing new openings for theoretical and empirical research which might reveal and quantify as yet undiscovered laws in ocean and atmospheric processes.[8]

Another area in which the need for global services is becoming apparent is discussed at some length in the Ponza Report, *Global Ocean Research.* The provision of a worldwide system to monitor pollution of the ocean could also provide

on a world basis an annual review of the state of the ocean and marine resources from the point of view of marine pollution and forecast long-term trends so that governments can take in advance the steps required to counteract its effects.[9]

As also mentioned in the Ponza Report, worldwide information concerning the ocean may best be provided through a general world science information system rather than a specialized one for the ocean, and this possibility must be taken into account. It would seem, however, that the unique characteristics of ocean data collection and dissemination would require particular handling for these data even though the information services were incorporated in a broader scheme.

In the fields of data and information management, the *Compre-*

hensive Outline of the Expanded Program contains the following recommendations:

The IOC Working Group on Oceanographic Data Exchange in collaboration with WMO, FAO and other interested organizations such as ICES should examine the above problems (of marine data and information management) and take requisite steps to meet the needs of the Expanded Program. Certain aspects of this work can be assisted by the IOC advisory bodies.

UNESCO, FAO and ICSU in collaboration with other interested organizations such as ICES should devote increased attention to the improvement of scientific information systems in the field of marine sciences.

Member States should give increased support to national, regional, and world data centres as required for the expansion and improvement of their services.

Specific mechanisms should be sought for accelerating the flow of data through international exchange channels. And, all meaningful data and information resulting from projects and programs of the Expanded Program should be considered as Declared National Programmes (DNP) or their equivalent, to be exchanged or available and subject to inventories.[10]

Little has been said thus far about instrumentation and methods or technology and supporting facilities and services, although these areas of service are important. In fact little has been done on a global basis under these subject headings. It would seem reasonable to assume that the development of global services in these areas might be similar to past developments in the areas of data handling. Once a suitable organizational framework is established for the development of global services, it could be expected that services in these newer areas would develop gradually, depending upon the demand for the various services.

Instrumentation is of immense importance to man's efforts to understand and make better use of the world ocean. Unless adequate instrumentation is available to provide reliable, accurate information about the marine environment, marine research, expeditions, and related efforts of the world community will have been wasted. Instruments, for example, must be properly calibrated if marine data is to be turned to effective national and international use.

In 1968, the United States established a National Oceanographic Instrumentation Center to

Operate a laboratory for the evaluation of oceanographic instruments

Generate a central proposal and specification file and disseminate information on ongoing development efforts for oceanographic instruments

Encourage the coordination of national specifications for oceanographic instrument development

Conduct cooperative programs among government agencies, the
academic laboratories, and the industrial community for the purpose
of assessing government-wide requirements on instruments to support
the development of standards

Establish techniques and secondary reference standards by which
oceanographic instrument performance can be assessed

Perform laboratory and field testing and calibration of oceanographic instruments for government, academic, and industrial interests

Collect and disseminate instrument performance and deterioration
data as a means of acquiring statistically significant samples on which
to base design criteria for improved systems

Develop ocean measurement instruments when these instruments
cannot be obtained from other sources and develop equipment needed
in the testing and calibration of oceanographic instruments

Little presently exists in the way of a much needed, comparable
global service. The *Comprehensive Outline of the Expanded Program*
states:

The Expanded Program will require the development and availability of instruments and methods of high precision and reliability. In order for data from various sources to be pooled and processed automatically, the instruments must be
intercalibrated or standardized where possible and methods must be compatible.

The following problems require solution:

—there is little effective intercalibration of measurements made by one
Member State, with any other Member State;

—information on the performance of instruments and related equipment is
not readily available to Member States;

—standards information to ensure high quality data is not available to Member States;

—information on appropriate facilities needed for the calibration of instruments is not available; no effective mechanism exists for standardizing on those
instruments which are worthy of such a designation.

The following actions should be taken:

—IOC, UNESCO, FAO, WMO, SCOR, ACMRR, ICES, and other interested
bodies should jointly intensify their support for methodological work and for
the improvement, intercalibration, and standardization of instruments and
methods.

—Member States should provide increased assistance in the conduct and publication of pertinent methodological investigations and encourage the production and adoption of standardized instrumentation where practical.

—Member States should designate, where possible, an existing laboratory or facility that can act as a Centre for information relative to that state's activities in oceanographic measurement and for the coordination of instrument improvement, calibration, and standardization with other Member States.[11]

Nearly all of the suggestions and recommendations concerning the needs for global services are similar in philosophy, if not in detail. By virtue of the responsibilities previously assigned to its individual components, NOAA is now the effective lead agency in the federal government for nearly all the global services discussed in this section.

With the assistance of the National Academy of Sciences and the National Academy of Engineering, NOAA should review the various national and international proposals for global monitoring and environmental forecasting in the atmosphere and ocean (including the biosphere) and for evaluating the effects of pollutants and should assess the feasibility of developing a comprehensive and unified system for these purposes. It should also assume the principal U.S. responsibility for promoting international data exchange and for strengthening charting and navigational aids for international use.

4

Regulation for Rational Use of the Ocean

TASKS REQUIRING INTERNATIONAL ACTION

It is increasingly apparent that important elements of the traditional legal framework for activities at sea are soon to be reassessed particularly in light of anticipated new uses of the ocean. There is emerging an insistent demand for a new look at the basic community policies at stake in use of the ocean (including serious questioning of the value of freedom of the seas in one or another of its various facets), at the principles of prescriptions regulating activity at sea, and, finally, at the procedures for implementing these policies and principles. Whether or not such an extensive reappraisal is needed (a subject on which considerable doubt is entertained by some observers, who believe a more intensive look at some specific problems would be more advantageous), the task must perforce be carried out on an international level, ranging from bilateral to regional to broad multilateral actions. Moreover, to the extent international activity is required, such activity must be deliberately concerted if a more rational use of the ocean and its resources is to be achieved.

For several reasons, international cooperative effort is needed to provide regulations for rational use of the ocean and its resources. As employed here, the term "rational use" refers to that combination of uses of the sea that provides optimal benefits to the widest possible group of states and peoples.

First, the ocean is global, but government, as the term is generally understood, is not. Whereas security of title and operations is provided where necessary by national authorities to facilitate use of land resources, international cooperation, in varying degrees, is required to facilitate the use of ocean resources beyond national jurisdiction.

Second, because of rapid advances in scientific discovery and technological capacity, the traditional uses of the ocean (and its resources) are intensifying while new uses are being invented.

Third, ocean resources are in increasing demand owing to the rapid expansion of the world's population (an expansion made possible by scientific discovery and technological innovation).

Both the increased demand in the number of uses of the ocean and the intensification of familiar usage threaten international conflicts for reasons originating in military, legal, economic, and political considerations. The interdependence of nation-states is nowhere more vividly depicted than in man's interactions in the ocean environment, and the need for international cooperation is nowhere more evident.

For these reasons, various, sometimes conflicting, national interests confront each other over ocean use and require continuous adjustment between and among states. In this respect, regulation of the use of the marine environment is far more difficult than regulation of resource use in outer space. The ocean has resources many peoples need, which they can and do exploit now. Compared with the regulation of outer space, ocean management presents a pressing political issue because nearly all nations can, or believe they can, utilize the ocean and its resources directly or by proxy. Moreover, the common-use tradition in ocean affairs, e.g., freedom of the seas, leads nations to affirm they have a right to do so. The technology gap becomes a pressing political issue in this context because developing countries, particularly those with little or no undersea capacity, fear they will be excluded from decision-making processes that might dispose of the economic and other benefits of two-thirds of the globe. The fact that many of these benefits may be derived from resources still beyond present national jurisdiction fortifies their claim to a voice in managing the world oceans. Inevitably, therefore, the exploitation of the ocean and its resources is the very stuff of international politics. The paragraphs that follow seek to account briefly for the intensely political character of international regulation of ocean use by mentioning some of the military, legal, economic, and political factors that shape the interests at stake.

MILITARY PROBLEMS

That military uses of the sea remain important in shaping the world balance of power is beyond dispute as the major powers employ the world ocean both in traditional ways and in radically new ways. Unclassified information amply indicates that the seabeds, including shelves, slopes, and abyss, are now being utilized for military purposes.

Although the current issue of *Jane's Fighting Ships* might imply that, with only two major naval powers remaining, the United States and the Soviet Union, the importance of naval power in world politics has diminished, the world's strategic balance depends in large measure on the ways in which these two nations exploit the military potential of the oceans. Major deterrent systems of the two superpowers are centered in the ocean, and both states locate surveillance and detection systems there.

This increasing significance of oceanic military uses has distinct importance for regulation of ocean uses generally. Since oceanic weapons systems are crucial components of the nuclear balance of forces, recommended international regulations for nonmilitary purposes will be examined critically for their impact on such military activities. For example, the specific authority to be exercised by an international seabed agency will certainly be assessed for its potential impact upon military operations and measures. Similarly, the determination of precise boundaries, as for the territorial sea and the continental shelf, will most certainly be appraised for effects on weapons systems and their use. These two illustrations suffice to establish that regulation for rational use is intensely affected by military considerations.

LEGAL PROBLEMS

Legal problems also contribute to the recent severe agitation about ocean development within international political bodies. Principal concern of states extends to two major categories of problems: the limits of national jurisdiction on the seabed and in the water column and provisions for gaining access to and sharing the benefits of resources located beyond such limits. Only concerted international action can dispose of these problems in an acceptable manner once some disposition is demanded. The converse of this statement is that no single state should be permitted to decide for itself on either of these questions without regard to the demands and legitimate expectations of the general community of states. Whether such unilateral determinations will in fact be made and permitted depends in part on the timing and character of international action.

With respect to boundary delimitations, states are exercised, some more than others, about the limit on the territorial sea, about the extent of exclusive rights of access to fisheries beyond the territorial sea, and about the limit to coastal control over the natural resources of the seabed beyond the territorial sea.

As recognized by the Convention on the Territorial Sea and Contiguous Zone, the sovereignty of each state extends beyond its land territory and internal waters "to a belt of sea adjacent to its coast" and "to the air space over the territorial sea as well as to its bed and subsoil."[12] Each state has the authority to apply its laws in its territorial sea and has permanent, exclusive access to the living and nonliving resources found in its internal waters or territorial sea, including the resources of the beds and subsoil.

Because there was insufficient agreement among nations in 1958 and 1960 at the Geneva Conferences, the Convention does not specify the breadth of the territorial sea. As of September 1, 1970, claims vary from 3 to more than 200 nautical miles, roughly as follows: The United States and twenty-seven other countries claim territorial seas of 3 miles; eleven countries claim 6 miles; forty countries claim 12 miles; eleven countries claim between 12 and 200 miles. The United States has recently expressed a readiness to negotiate a 12-mile limit, if certain conditions concerning transit rights are also adopted.

Since 1960, in the absence of explicit international agreement, states have gradually, and in a few instances markedly, extended the limit of their territorial sea. Sometimes force is threatened in order to sustain a unilaterally determined limit. This continuing extension of national control to the potential detriment of freedom of access for varying purposes, including transit and overflight, is responsible for the recent urgent suggestions for a conference to negotiate specific limits on this and other national boundaries.

This uncertainty about territorial sea limits also attends the question of fishing limits. Numerous states now claim fishing limits beyond their territorial sea, including the United States, but there is limited explicit agreement on how far beyond is proper. Some states claim many more than 12 miles, a few claim less.

Finally, regarding boundaries of national authority in the ocean, a pressing political issue is the limit of the continental shelf. The definition in the existing Convention on the Continental Shelf of 1958 is not precise, and, in December 1970 at its 25th session, the UN General Assembly decided to convene a conference, planned for 1973, that will consider this as well as other issues. Technological advance has virtually erased the depth definition of 200 meters as an effective limit and has underscored instead the criteria of adjacency and exploitability. The legal argumentation concerning the proper interpretation of these terms has been lengthy if not illuminating, ranging from placing boundaries in midocean, to broadening shelf limits to include the entire continental margin as part of the shelf, to adopting very restrictive conceptions. In the Presidential statement of May 23, 1970, the United States observed that the matter is far too important for solution by doubtful legal interpretations and suggested a new treaty providing for a 200-meter shelf limit and an international regime beyond.

Although overall political assessment may suggest revision of the Continental Shelf Convention, there is still something to be said for the present legal regime resting on the current wording of the Continental Shelf Convention. The exploitability clause serves to link technological capacity with exclusive rights in the areas adjacent to the coast. If "adjacency" can be interpreted as extending to the edge of the continental margin, nations can exploit as far seaward as engineering capacities are apt to take them during any period practical to plan for, since economic exploitation on a significant scale beyond the continental margin is believed to be more than a decade away. For the immediate and middle-range future, the present international regime encourages national exploitation because it provides security of claims protected by known national authorities. A revision of the present regime would probably be as hard to agree upon as was the one accepted in 1958. A new type of regime might delay or impede exploitation by lessening the security of property rights. Furthermore, it can be argued that any immediate revision should be very modest until the availability of resources is better known as a consequence of scientific exploration and commercial pilot-scale operations.

There are weaknesses in the present international regime, however, because technological advance promotes multiple uses of the oceans. If the extraction of oil and other minerals were the only capacity at stake, the present regime might well be sufficient. In fact, however, the same technological skills that make minerals available also provide opportunities for communications and other military uses of immediate concern to coastal nations. The present regime permitting exclusive national claims to expand seaward may not be consistent with some national security interests. Commercial and other considerations may argue for expanding territorial claims or wider exclusive jurisdiction over resources, while military exigencies may call for restrictive national claims. A closely related weakness is that some nations will make extreme claims to exclusive national jurisdiction as a means of protecting future interests unless there is an effective moratorium on the extension of national claims or agreement to limit such claims in specific ways.

Controversy abounds regarding both the basic policies at stake in exploitation of "common" resources beyond national jurisdictional limits and the principles to govern use of this area, including the accommodation of potentially inconsistent or conflicting uses. Failure to reach agreement on fundamental policies and at least guiding principles is thought by some to inhibit resource uses. At present, states continue to experience major difficulty in reaching consensus on even the broadest principles to govern this area.

Agreement on a legal regime could become more difficult as the scope of ocean activities increases. If a primary objective of national policy is to increase the national income of the coastal states by the extraction of seabed minerals, present arrangements will doubtless serve. If, however, national policy emphasizes other objectives such as arms control, increased funds for developing countries, including those not contiguous to the sea, and maximizing the development and rational management of fisheries resources, the regime needs overhauling. There is, in addition, considerable advantage to be gained from an agreement allocating some limited authority to an international body over mineral development and other uses so as to preclude any continued felt need for expanding national controls from one activity to another as states perceive possible harm from an activity uncontrolled by any political authority. It will similarly be desirable for any international agreement to assure that such international body as may be created does not similarly seek or assert continued expansion of its control except as may be provided in its basic charter.

Whatever the outcome of these arguments about coastal limits and the regime beyond, they require settlement by international agreement. A danger is that, in seeking an immediate international conference on these problems, solutions will emerge that defer excessively to coastal state interests at the expense of more general interests. Another possible difficulty is that provisions acceptable to a large enough number of states cannot be negotiated, leaving the situation as fluid as before the conference and inviting, more likely than not, unilateral extensions of jurisdiction for many or all purposes.

ECONOMIC PROBLEMS

A major question rooted in the economics of ocean resources is whether increasing demands and capabilities call for more fully developed international regulatory mechanisms. A further question is whether such mechanisms should be separate special-purpose regimes for living and nonliving resources, respectively, as at present, or whether there should be an overall general-purpose international ocean agency to regulate the uses of the ocean as a whole. To date, separate mechanisms are thought to have been adequate and there is little to suggest that an overall agency would be more effective as ocean usage and, therefore, management become more complicated. The answer in the future may be influenced by the extent to which national jurisdiction moves seaward and by whether it moves farther out and faster on the seabed than in the water column or on the surface above. The exploitation of North Sea mineral and living resources suggests that, under circumstances unambiguously involving the continental shelf and involving states experienced in cooperative activities, one type of jurisdiction (limited sovereign rights) can be operated successfully on the seabed with another (high-seas common property) in the water above. Whatever specific regimes become necessary, however, it seems inescapable that increased economic benefit from ocean resources exploitation requires more highly developed forms of international cooperation.

Economic considerations extend beyond the distribution of resources or of the benefits of their exploitation to influence governmental attitudes toward international scientific cooperation. This is seen most clearly in the views of the developing states toward the role of the Intergovernment Oceanographic Commission in promoting marine science research. The report of the U.S. delegation to the eleventh meeting of the IOC Bureau and Consultative Council in January 1970 offers a succinct view of the very different perspectives of developed and developing states toward cooperation in marine science:

The developed nations favor broadening the scope of the IOC's work along the lines of the proposed Long-Term and Expanded Program of Oceanic Exploration and Research (LEPOR). These nations favor strengthening the IOC to the extent necessary to carry out this broadened responsibility. The developed nations generally believe that such exploration and research is not tied to exploitation of resources and can be conducted anywhere anytime under the existing freedom of scientific research and the existing freedom of the high seas for the benefit of all mankind.

The developing nations also favor strengthening the IOC, but not primarily for the implementation of LEPOR. Their intent is to build a stronger IOC for the purpose of assisting the developing nations to exploit their off-shore resources and to establish a strong international regime for directing the development of deep-sea resources for the benefit of the developing nations. Thus, the role that the developing nations envisage for the IOC is an increasingly restrictive and regulatory role which would carefully direct scientific exploration and research towards the development of resources, rather than encourage the free development of non-resource-oriented research. As a corollary to such restriction, some of the developing nations refuse to recognize the freedom of scientific research and the freedom of the high seas.[13]

Further indication of the influence of economic considerations is seen in the reception accorded the International Decade of Ocean Exploration. When the United States proposed the Decade at the United Nations, the developing countries were hardly convinced that the overriding priority was careful study of the marine environment. Many were more concerned with the economic consequences of scientific investigation. Although the U.S. proposal was cosponsored by approximately thirty UN Members and was adopted by consensus (i.e., "without a vote"), there were undercurrents of discontent and skepticism.

Although all coastal nations can facilitate the Decade's achievement by permitting investigation in areas under their jurisdictions, only a handful of technologically advanced nations are equipped to carry out much of the work of the Decade in any major way. Under these circumstances, the suspicion arises that international cooperation in ocean investigation will redound primarily to the economic advantage of the rich nations, or at least will do nothing to provide any special advantages for the poorer nations. Some governments fear that instead of closing the income gap, or braking its rate of increase, the exploitation of ocean resources will widen it unless discriminatory countermeasures are taken. With these considerations in mind, apparently, several members of the UN Sea-Bed Committee have stated that scientific research must not be allowed to establish exploitation rights on the seabed beyond the limits of national juridsiction. Numerous countries, including those that are developed, urge that LEPOR should include concrete plans for strengthening the research capabilities of developing nations. Technical assistance would thus appear to be an important catalyst in gaining international cooperation for the Decade. The latter, in turn, will be attractive to developing and developed countries alike, largely to the extent it promises economic rewards.

The case is similar with respect to international arrangements for the exploitation of nonliving resources. Expectations or hope of economic benefits from mineral resources on the shelf or in the depths beyond tempt all nations, whether industrially advanced or industrially backward, to expand their claims of exclusive jurisdiction over seabed resources. Any counter proposals to establish international regulatory bodies are scrutinized carefully in terms of economic payoffs. The question is whether a more fully developed international regime than now exists holds promise of economic rewards for rich and poor nations and for coastal and noncoastal nations. The noncoastal nations also have a stake in the ocean environment as the common heritage of mankind.

The case with living resources exploitation is also similar. Changes in regimes and the development of new international mechanisms must hold promise of improved economic benefits. Some developing countries, in particular, are becoming increasingly dependent on fisheries for food and foreign exchange. Many have taken steps to increase their rates of growth in fish production above that of the developed countries. Some of them, moreover, are relative newcomers to membership in international fisheries bodies. All of them are concerned that international regulatory mechanisms might work to their disadvantage as they compete with the rich and with each other for scarce resources.

International agreements regulating living resources exploitation are improving. As readily exploitable fish stocks are threatened by improved fishing techniques that enable an increasing number of nations to make bigger catches far from home ports, and as a hungry world presses against available food resources, governments are beginning to seriously explore allocation of resources that have been exploited traditionally under the "common use" principle. They sometimes seek to agree, at least *de facto*, on how much each nation should take of a particular stock. Some governments are also beginning to take definite steps to establish restrictions on fishery effort through international conservation agreements monitored by international fisheries commissions, as by tacit agreement on differential access to particular stocks or by limiting the number of vessels permitted in a fishery.

The most obvious short-term result of rapidly advancing fishery technology is international disagreement, extending on infrequent occasion to violence. Steps to assert rights for living resources that are comparable to property rights have been taken in some parts of the world without international agreement. Most notably, some Latin American countries have gone beyond conservation purposes to exact license fees from foreign fishermen who wish to exploit resources within 200 miles of their coastlines.

POLITICAL PROBLEMS AND FOREIGN POLICY

Foreign policy for the oceans is intended to promote national interests that are variously defined in terms of preferred patterns of ocean use. The emphasis in the present discussion has been upon military and extractive uses, but other important activities with political implications include transportation, recreation, waste disposal and pollution control, and scientific investigation. Each of these is considered in calculations of national interests. A nation must perforce take account of how its pursuit of a particular ocean use affects its relations with other nations and other participants in use.

The principal political factors that contribute to the drive for international action are: (1) the growing disparity in affluence between the developing and developed countries; (2) the sudden increase in the number of independent states, almost all of which are poor and feel they should have a voice in how ocean resources are to be used; (3) the growing congruence between the United States and the Soviet Union in certain aspects of ocean affairs (e.g., military uses of the ocean, the freedom to conduct scientific research at sea, and restraint in extension of national boundaries) that threatens to line them against other countries, particularly those less developed; and (4) a widespread expectation that regulating the uses of the oceans requires institutionalized multilateral arrangements.

On the first of these factors, anticipations of wealth from the exploitation of living and nonliving resources, whether warranted or not, have excited the imagination of rich and poor countries alike. While some nations seek to halt the rate of increase in the growing income gap by extending claims to national jurisdiction or by limiting such jurisdiction in favor of international authorities, they are confronted with the task of forging new international agreements to achieve their aims. For this reason, many developing countries look to international action such as a new general law of the sea conference, in the hope of increasing their economic benefits from ocean exploitation. Others, particularly the United States and other nations advanced in ocean capabilities, prefer to move step by step, first seeking international agreements on issues still unresolved, such as the territorial sea limit, transit rights, fishing limits and rights, the shelf limit, pollution control, and the regime beyond the limit of national jurisdiction. A conference on the entire span of problems in ocean use is thought by the United States and other states to jeopardize agreement on specific problems in more limited areas. The less advanced countries favoring a general conference are fearful that the advanced countries will widen the income gap by pre-empting the lion's share of ocean resources. In their view, a general conference provides them greater opportunity for bargaining.

In making decisions on ocean regulation, it is no longer possible for a few major maritime powers to determine and enforce the law of the sea. They no longer control vast overseas areas that, by virtue of colonial subordination, formerly conformed to agreements reached among the privileged few. The limit to the number of independent sovereignties has not yet been reached. The new nations are well aware they can gain a measure of leverage (sometimes substantial) against the older, richer nations by concerted political action in the United Nations and other international bodies. However, opportunities to use international conferences for national ends are by no means limited to the less advanced new nations, despite the advantages they gain in equal voting rights in many, but not all, such bodies. The major powers, in competing with each other, are also often tempted to use conference diplomacy to curry favor with the developing countries.

Despite this temptation, the Soviet Union and the United States have displayed a considerable congruence of interests in ocean affairs. As major oceanographic powers they are targets of demands by less advanced countries to reach agreement on economic matters with the lesser powers, but their greater concern is to settle problems, chiefly military, of direct consequence to themselves. The effect is to supplant the tensions of an idealogical (East–West) struggle with those of an economic (North–South) confrontation. While this interpretation can be carried too far, the United States and the Soviet Union have quite consistently opposed capital development funds in the United Nations, resorting to bilateral development assistance far more than multilateral development assistance. Both the United States and the Soviet Union have come under attack in the UN Conference on Trade and Development. Nonetheless, they differ considerably with each other in their support of international organization. Because of success in promoting some international institutions, such as the World Bank and the UNDP, that it can very largely control by weighted voting and special representational arrangements, the United States, on balance, looks on international institutions with greater favor. A major exception to the Soviet Union's view of international organization is its strong backing of UNESCO's Intergovernmental Oceanographic Commission, in which it is now nearly the equal of the United States as an oceanographic power. In any event, less-advanced oceanographic powers are taking international action to strengthen their hands against the superpowers.

Finally, there is an almost irresistible temptation to seek multilateral solutions to the problems arising out of increased usage of ocean resources. The temptation is understandable. The UN and many other multilateral organizations, useful in mobilizing political support for national interests, were in existence before the management of ocean space as a whole became a pressing international political issue. Since ocean space is still mostly unclaimed by national authorities, there is a *prima facie* inclination to promote international action as the means of providing for the governance of activities on and under the sea. The world ocean is the stage for natural events and processes of immense, sometimes global, proportions, requiring investigative and other actions on a similar scale. In short, multilateral techniques seem desirable as elements of a stable public order in the oceans. Indeed, the pressing need for international cooperation in managing the global environment implies the need to consider international institutions as *objectives* of foreign policy as well as *instruments.*

Whether international organizations are viewed by governments as goals or instrumentalities, it is clear that the present UN system as a whole is ill-equipped to conduct the global operations that the management of global ocean systems would seem to require. The political processes of the General Assembly do not ally political influence with operational capacity in an effective manner. The decentralized system of specialized agencies and non-UN bodies is presently too loose. Confidence in the present structure of international agencies is conspicuously lacking in both the advanced and less advanced countries. Multilateralism generally arouses little confidence in the United States (and less in the Soviet Union). It is equally clear that, without improved means of intergovernmental collaboration, there is a propensity toward unproductive conflict that could militate against man's growing technological capacity to use ocean resources.

MECHANISMS FOR INTERNATIONAL ACTION

NATIONAL MECHANISMS

Because of the number of interests and agencies involved in the United States, the national decision-making process for ocean policy is diffused throughout much of federal and state government. The federal government, moreover, is characterized by organizational pluralism in both the legislative and executive branches, reflecting the broad gamut of interests involved in ocean affairs, including national defense, petroleum extractions, fisheries, transportation, and recreation. The task of administration and organization is to facilitate a decision-making process capable of harmonizing particular interests and programs in support of a broad national policy for ocean use. The fact that the ocean and its resources mostly lie beyond traditional limits of national jurisdiction, plus the fact that exploitation *within* national limits affects other nations, leads inescapably to foreign policy and the Department of State.

The Department of State   Despite the growing importance of the White House Office and its National Security Council Staff in coordinating many foreign affairs activities of the executive branch, the Secretary of State is still formally expected to be the President's principal foreign policy adviser. He also remains charged with the actual conduct of such daily foreign policy affairs as are managed by his department.

Responsibility for formulating and implementing ocean policy within the Department is diffused, much as it is in the executive branch, among a number of geographic and functional units.* The Special Assistant to the Secretary for Fish and Wildlife is now the official responsible for the increased activity of the Intergovernmental Oceanographic Commission and regularly conducts relations with other states and international organizations concerning fisheries of

*See footnote on p. 15.

mutual interest. Previously, the U.S. Mission to the UN had taken the lead in representing the United States in the General Assembly and in its Sea-Bed Committee, although more recently this function has been assigned to the Legal Adviser of the State Department. The Legal Adviser also has primary responsibility in negotiating at conferences on the law of the sea. The Bureau of International Organization Affairs is involved in ocean policy to some extent because of the international agencies concerned: UN, UNESCO (IOC), FAO, IMCO, or WMO. Regional organizations concerned with certain aspects of ocean policy, such as NATO, the Organization for Economic Cooperation and Development, and the Organization of American States, are backstopped largely by the Bureaus of European Affairs, Economic Affairs, and Latin American Affairs. Still another officer of the State Department, the Director of International Scientific and Technological Affairs, is involved with marine science matters, including, for example, policy issues concerned with the International Decade of Ocean Exploration and environmental issues.

With responsibility so diffused, there may be a question whether the State Department is organized to advise the president coherently and consistently on the overall foreign policy dimensions of ocean policy generally and marine science particularly. A further complication is the temptation of U.S. missions, at the UN and other international bodies, to take inadequately coordinated policy initiatives.

One question is the extent to which international marine science affairs should involve the State Department. The Department's leadership in representing the United States in international conferences dealing with oceanographic matters is viewed by some scientists as a symptom of an undue political influence on marine science. The first recorded vote of the IOC (recorded at the 1969 Paris meeting) concerned freedom of scientific inquiry—a matter of great importance to oceanographers. While representation by government officials was greater than at previous IOC conferences, it will be important to ensure that, in future, there is a strong representation of the scientific community. However, since the promotion of marine science often requires diplomatic action, continuing State Department involvement to the point of leadership on some issues in ocean affairs seems inevitable.

The Executive Branch A more significant measure of the difficulty of coordinating ocean policy is the array of departments and agencies with which the State Department must deal. The most important are involved with military and commercial uses of the ocean and with monitoring the global environment. These uses and functions increasingly take place beyond national jurisdiction and involve other countries, whether the activities concerned take place on the high seas or in nationally claimed areas. In marine science affairs, the principal operating and funding departments and agencies include the Defense Department, the National Oceanic and Atmospheric Administration in the Commerce Department, the National Science Foundation, and the Coast Guard, plus HEW and the AEC. The Agriculture and Labor Departments also represent interests involved with marine science and resources. Nearly all noncoastal oceanographic research in academic laboratories is funded by the National Science Foundation and the Office of the Naval Research.

Coordination in the executive branch is sought at several levels by many means. As noted previously, most important is the cabinet-level National Council on Marine Resources and Engineering Development, with the Vice President as Chairman, established by Act of Congress in legislative recognition of the need for a national ocean policy. Below the cabinet level, coordination of some international components of federal oceanography programs is sought by a Committee on International Policy in the Marine Environment, established by the State Department and consisting of various departmental representatives but acting chiefly through a working-level interagency Panel on International Programs and International Cooperation in Oceanography.

INTERNATIONAL MECHANISMS

The principal organized method for prescribing international law is through multilateral conferences organized by the United Nations. Numerous other international bodies also participate in decision-making in connection with specialized ocean uses such as military activities, living resources, transportation, research, and communications. Some of these bodies are specialized agencies of the United Nations, but others, particularly those concerned with fisheries, are not.

Historically, states have acted to create law by the unorganized method of unilateral claim and response exhibiting a pattern of behavior that creates the expectation that certain conduct is lawful. With the proliferation of states since World War II and the increasing complexity of policy issues, this method obviously leaves much to be desired.

The United Nations Involvement of the UN, principally the General Assembly and International Court of Justice, began shortly after World War II. The Court has made a number of its most important contributions to clarification and creation of international law by its pronouncements in cases arising from disputes over ocean regions. The *Corfu Channel* case and the *Anglo-Norwegian Fisheries* case, decided in 1949 and 1951 respectively, are widely regarded as significant pronouncements having marked influence far beyond their immediate occasion. The recent (1969) decision by the Court in the *North Sea Continental Shelf* cases may also have similar import.

The General Assembly quickly became associated with creation of law for the sea, principally through the activities of the International Law Commission. Beginning in 1949 and continuing for several years, the Commission laid the basis for the conferences in Geneva in 1958 and 1960 that produced the four conventions on the law of the sea. These treaties contain virtually all the customary law then applicable to the ocean and create some highly important new law for the sea. The General Assembly's role in this process was primarily that of arranging and sponsoring multilateral meetings that served as negotiating arenas. The primary background work that served as the bases for discussion came from the International Law Commission, which on fisheries questions had the very substantial assistance of a preliminary FAO conference on certain technical matters.

During the decade of the 1960's, the nature of the UN involvement in the decision process changed substantially. In contrast to earlier years, the General Assembly's role is now much more active than merely serving as principal sponsor for conferences to adopt treaties on the law of the sea. The principal activities of the Assembly include focusing attention on some problems that appear to pose imminent possibilities of conflict (unless the community of states provides a better legal framework) instigating the collection of information about the needs for scientific exploration of the sea, and providing a negotiating forum for arranging the agenda and timing of a new conference on the law of the sea. The Assembly has also provided a means for negotiating about some of the fundamental principles applicable to some uses of the sea beyond national jurisdiction.

The significance attached to these Assembly activities is evidenced both by the assignment of agenda items concerning the oceans to the First (Political) Committee rather than the Sixth (Legal) and by the creation of a Standing Committee on Peaceful Uses of the Seabed Beyond the Limits of National Jurisdiction.

There can be no doubt that the United Nations, and particularly the General Assembly, will continue to play a role in creating policy and law for some important ocean matters. The current UN involvement will certainly produce a new multilateral conference, and perhaps a series of meetings over a period of years, to attempt resolution of some pressing problems. In the absence of the UN structure, the convening of such gatherings would probably be very difficult to achieve and would certainly take a much longer time to arrange. In these terms, the UN contribution to an orderly process of deciding disputes about ocean use cannot be overestimated. It remains to be seen, however, whether or not the decision process will in fact resolve the problems dealt with.

The Intergovernmental Oceanographic Commission   The actual coordination of specialized national and international programs need not involve the all-purpose General Assembly or the United Nations itself. When activities such as the International Indian Ocean Expedition are involved, special-purpose bodies with membership and operating procedures suitable for the operations' objectives have been accepted. Within the existing structure of international agencies, the IOC is the prime candidate to be a global ocean agency. It could operate in ocean affairs much as the World Bank operates in economic affairs or the International Atomic Energy Agency in nuclear affairs. In several functional "agencies," operating capacity has been linked effectively with policy control by extending special rights and responsibilities to the advanced nations.

The IOC is the prime candidate for a global marine science agency for several reasons. Its membership is increasing, though slowly. Its executive direction through its Executive Council has been scientifically competent and, for the responsibilities so far assigned, administratively efficient. It has developed stronger cooperative links with other ocean agencies, notably FAO, The World Meteorological Organization, and UNESCO, its parent organization, by the establishment of an Inter-Secretariat Committee on Scientific Programs related to Oceanography, composed of the executive heads of these agencies. Other agencies may be invited to join this group if they express willingness to contribute personnel to IOC's secretariat, to coordinate their programs with IOC programs, and to seek IOC's advice in marine science affairs. Its own professional staff is being doubled. These steps have been taken to implement the Long-Term and Expanded Program in Oceanic Research and Exploration endorsed by the General Assembly. Of course, whether they are adequate for that purpose remains to be seen.

The IOC is also a forum for political persuasion, sharing this function with other bodies of the UN and the Specialized Agencies. It is a political forum to the extent that its members seek to advance their policies on such matters as the freedom of scientific inquiry or the substantive content of the Long-Term and Expanded Program of Oceanic Research and Exploration.

The major difficulty rests not so much on the doorsteps of the international agencies as on those of the member nations. An international oceanographic program can be no greater than the resources committed to it by national governments. At the policy level, the agencies are important for the mobilization of political support. At the operating level, they can enhance the impact of national ocean programs by international coordination.

A second difficulty is that the IOC must compete for financial and administrative support in UNESCO and in other bodies whose attention is at best focused only partially on marine science. However, given the reluctance of advanced countries to create new agencies whose activities they are expected to finance, the present efforts to strengthen IOC are indeed necessary. If LEPOR runs into difficulty, the advantages of a broadened treaty-level marine affairs agency may become more apparent.

Fisheries Commissions Over twenty fisheries commissions have been established to perform certain functions concerning the conservation of living resources of the sea. Through its Fisheries Department, Committee on Fisheries, and advisory committee, FAO also is a major element in the institutional structure for world fishery management, although it does not itself engage in regulation.

From a global perspective, a noticeable weakness of the existing system is its *ad hoc* nature. Created largely to deal with certain specific problems as they were perceived, the commissions apply only to selected species and regions and too often reflect only transitory patterns of national fishing involvement and interest. There is no formally organized global system or regional fishery commission to oversee management of fishery resources generally, nor has this task been undertaken by any international organization in a systematic way. Some work in this direction has been done by the FAO Committee on Fisheries, but the level of attention directed to this problem is not adequate.

The major reason for the inadequacy of the fishing commissions is the persistence with which the member states retain their sovereign prerogatives. This is manifested in a number of important ways. First, and most important, states have been most unwilling to create management entities that have any significant independent authority. Most of the existing commissions are little more than continuing negotiating arenas within which members may bargain with each other to reach mutually acceptable solutions. If such solutions can be negotiated, then the commission can make recommendations that members may implement. If, however, these negotiations do not meet with success, the commission cannot itself act. In short, the commissions are seldom distinguishable from the constituent member states as a group of still-independent sovereign states. Some progress has been made toward changing this situation, notably in the northwest Atlantic, but the difficulty persists in most of the commissions.

Further illustration of the retention of authority in state hands is seen in the niggardly funding of international fishery commissions. On a worldwide basis, an almost infinitesimal fraction of the value of the world fish catch is devoted to management expenditures. Obviously the international commissions are not likely to be effective managers if not granted the resources needed for this task. This denial of funds to international agencies in itself might not be a serious defect if expenditures by state agencies were adequate. However, in the United States, investment in management by national agencies is declining rather than increasing, and there does not appear to be any real prospect of improvement. The responsibility for this situation rests primarily on budgetary authorities within the federal government. The low level of expenditures by the United States for international institutions is partly attributable to the congressional committee that oversees State Department appropriations.

In several other respects, too, the commission system is deficient. The objectives sought by the several commissions strongly resemble one another, for each commission adopts a slightly different version of the maximum sustainable yield as the goal of management. It is widely understood that agreement on this objective represents the lowest common denominator acceptable to the states concerned. With some exceptions, mostly in the northwest and northeast Pacific, it is only recently that fishery commissions have been charged with dealing with the more crucial problem of determining who gets what from the common resource.

A major weakness of the fishing commissions is the prolonged time usually required for dealing with anticipated, or actual, management problems. The vast fishing capacity now exerted by highly mobile fleets places such pressure on fisheries resources that problems can emerge very rapidly and place a corresponding strain on the management structure to cope with them. Major improvements in the system are required to deal with this situation.

For the most part, the fishery commissions are not even authorized to develop, by research with their own resources, the basic data necessary for successful management. Only three commissions are permitted to employ an independent research staff, and all of these were created over two decades ago. All other commissions rely on investigations by national agencies and thus devote major efforts to coordinating national research activities. There are no signs that any state desires to create further international research staffs, although it is widely acknowledged that the three existing staffs have been extremely useful and effective. The commissions served by these staffs (International Pacific Salmon Fisheries Commission, International Pacific Halibut Commission, and Inter-American Tropical Tuna Commission) are also probably the most successful in attaining the objectives set for them.

RECOMMENDATIONS CONCERNING U.S. POLICY

The preceding chapter examined recommendations concerning mechanisms for marine research. We now turn to the broader question of mechanisms for national and international actions to formulate policies, principles, and procedures for rational use of the ocean. Some of the recommendations already made for improvements apply in substantial measure to mechanisms directed at this more general task of decision-making about ocean interactions. However, this task, concerned with the whole panorama of ocean use, involves not only a broader subject matter than marine research alone, but also a complex set of functions that must be performed both on national and international levels. Accordingly, the observations that follow build upon and supplement those made in the preceding section. The recommendations are made in terms of the more important functions that need to be performed and the institutional innovations or rearrangements that are needed in order to do the job.

NATIONAL MECHANISMS

Intelligence Function   At present, there is no single central entity that acquires the data relevant for rational decision-making about ocean use. This is not merely a question of reorganizing the structure of government by amalgamating certain agencies or devising new coordinating mechanisms. Decisions regarding ocean policy must be based on information concerning important trends, including (but not limited to) development of our knowledge of the ocean environment. Such information must be consciously sought, compiled, and prepared for dissemination to users (decision-makers). An office or group within the Department of State should be designated to develop both the information base that is required and a means for feeding the data to those who need it. It is plain that execution of this function is a complex task and that the resources needed for its proper discharge are not minor.

A beginning has been made in the development of a somewhat similar system, centered within the Department of State, but the function is too narrowly conceived and the system grossly underfunded. The conception is too narrow because it aims only at an anticipated new law-of-the-sea conference, whereas U.S. foreign policy in the ocean involves vastly more than this, including an enormous network of bilateral and limited multilateral relationships. Resources are being drawn from existing personnel and resources when the need is for additional resources.

Initiation of New Policies Within the United States, the processes by which authoritative choices are made are initiated by a number of agencies and groups, often without regard to the decisions of each other. In terms of bilateral relationships, this function would involve making a proposal to negotiate or making a response to such a proposal by another state or making a decision to institute a suit in an international forum. With respect to multilateral relationships, i.e., organizations, this function would entail placing an item on the agenda for general discussion and disposition by the group. In both of these senses the decision-making authority is very widely dispersed in the U.S. Executive Branch.

An overall U.S. policy relating to the ocean (vis-à-vis other states and groups of states) can probably be achieved only by placing one official in charge of policy supervision. Although it would be feasible to continue to make policy on a somewhat decentralized basis, there should be one office charged with ensuring that an overall U.S. policy is established and adhered to. The purpose of this consideration is to reinforce the belief that a central office in charge of ocean policy should be established in the Department of State.

Formulation of Ocean Policies The United States participates in making policy for the ocean both by its unilateral actions and by joining with other states to make agreements regulating their individual actions. As noted, the responsibility for deciding what the United States should do itself and what agreements it should seek is greatly decentralized.

Within the State Department, a single office should be assigned responsibility for developing a coherent ocean policy, for maintaining U.S. positions consistent with that policy in the various intergovernmental bodies, and for coordinating and supervising pertinent activities of offices within the Department concerned with ocean affairs. This office should also compile, analyze and promulgate information required for rational decision-making on ocean problems. *

As the Secretary's principal staff aid in the conduct of ocean diplomacy, the Special Assistant for Ocean Affairs would coordinate and direct the activities of the several geographic and functional bureaus and offices that are already, or should be, fulfilling advisory roles in framing ocean policy. The new post would be an extension of the present post in the Secretary's Office of Special Assistant for Fisheries and Wildlife, expanded to include such components of ocean affairs as marine science and technology represented in the Department's Bureau of International Scientific and Technological Affairs.

Duplication of the reporting, representational, and policy-implementing functions of geographic and functional bureaus already involved in economic, political, and legal aspects of ocean affairs should be avoided. The position would be a staff aid to enable the Secretary to establish and coordinate a coherent foreign policy for particular ocean activities rather than a line bureau competing with, or replacing, existing units.

Equally important to intradepartmental coordination would be the Special Assistant's contribution to interdepartmental coordination. He would be free and qualified to assist the Secretary in dealing with departments and agencies (Navy and Commerce, for example) conducting major ocean programs. With regrouping of agencies to form the National Oceanic and Atmospheric Administration, the Special Assistant would be useful in advising the Secretary of State on the foreign policy aspects of the new agency's ongoing or proposed ocean programs. Depending on the level at which the office is estab-

*See footnote on p. 15.

lished, its personnel might represent the Secretary of State on CIPME and on working groups such as PIPICO.

Congressional liaison would be an important function for the Special Assistant. He could play a useful role in coordinating the State Department's relationships with several congressional committees. The proposed post would assist in both intradepartmental and interdepartmental coordination in seeking for the President supportive action by Congress.

The new federal ocean agency, NOAA, does not encompass all major federal programs, since substantial activities are still conducted by other agencies, including the Navy, NSF, and the Coast Guard. Accordingly, there is still a need for coordination of these other federal programs with those now in NOAA in order to formulate a coherent national ocean policy and to cooperate with other governments in international programs. *For the purpose of coordinating activities of NOAA with the ocean programs of the Department of Health, Education, and Welfare, the Atomic Energy Commission, the National Science Foundation, the Department of Defense, and other federal agencies, an interagency committee on ocean affairs should be established under the chairmanship of the Administrator of NOAA.* A new interagency committee on oceanography chaired by the NOAA Administrator and composed of high-level officials from the various agencies would serve as a satisfactory means for meshing the various ocean programs as well as give them the kind of visibility required to secure the necessary resources in support of the various programs.

The regrouping of federal ocean-oriented agencies in NOAA should be reinforced by rationalization of relevant committee structure in Congress. Congress should review the jurisdictions of its legislative and appropriations committees in the light of division of responsibilities for marine activities that may result from reorganization within the Executive Branch. According to the Stratton Commission, some adjustment will be necessary in the jurisdiction of congressional committees to achieve a coherent national focus for marine activities. The Environmental Science Services Administration had difficulty in achieving a balanced program because of the necessity of reporting to three separate legislative committees in the House. The objective of congressional reorganization would be to place the activities of NOAA within the purview of a single legislative and appropriation committee in each house.

All marine fisheries subject to U.S. jurisdiction should be regu-

lated by an appropriate federal agency, at present the National Marine Fisheries Service.

Historically, the regulation of fisheries in U.S. waters has been vested in the individual states, the role of the federal government being primarily to conduct research and provide funds to the states. There is increasing recognition that this allocation of regulatory authority to the states is a major impediment to restoring the depressed segments of the U.S. fish-catching industry. The Stratton Commission Report and the report of its Panel on Marine Resources provide a persuasive demonstration of the inadequancies of state regulation. For the reasons cited by the Stratton Commission, and because we have seen no attempt at improvement in the situation in the two years since the Stratton Report was issued, the Panel recommends that the National Marine Fisheries Service be authorized to regulate all fisheries subject to U.S. jurisdiction.

Some specific fisheries, such as those for sedentary species (e.g., clams and oysters) are local in nature. Regulation of these fisheries might be delegated by the federal agency to an appropriate local body.

Appraisal of Effectiveness of Ocean Policy It is astonishing how seldom agencies consciously and systematically assess the results of prior and current decision-making to determine whether objectives or policy goals are being served as hoped, and a special group is almost never created for just this purpose. At the very least, every agency with any decision-making function with regard to the ocean should deliberately appraise the consequences of its own actions. However, in view of the general tendency to refrain from conceding error, it would seem most advantageous if the appraisal process were in the hands of a special group other than the one whose actions are subject to examination. *The systematic appraisal of the effect of decision on the achievement of ocean policy objectives should be carried out by an appropriate federal body.* Within the United States, the actions of the executive branch might appropriately be kept under surveillance for their policy effect by a centralized office either in the Department of State or in the Executive Office of the President. The latter seems the better choice in view of the several agencies involved in ocean policy-making besides the Department of State, but selection of the appropriate agency or creation of a new body for this purpose requires careful consideration.

In the immediately previous section, the need for a central inter-governmental agency for ocean science and engineering was recognized, and reference was made to its possible components. Creation of such an agency could provide a specialized input to those international bodies considering political, economic, and legal aspects of ocean affairs, serving a similar intelligence function on the international level to that envisaged for NOAA on the national level. This task would be an enlargement of that now foreseen for IOC in its responsibility for the Long-Term and Expanded Program in Oceanic Research. The knowledge and information generated by this latter effort may provide a firmer base than now exists for a regime of deep seabed exploration and exploitation and also for regulation of other uses in areas beyond national jurisdiction.

It is, however, to be noted that international agencies are in some respects more functionally oriented presently than are national agencies. Thus, specialized consideration is given by different bodies to fisheries matters, military questions, and seabed development, in addition to the particular attention devoted to marine science and engineering by the IOC. Amalgamation of all these bodies into one centralized institution may not be necessary, but clearly some improvement must be sought in existing bodies.

A great many improvements are urgently needed in the international fishery agencies if they are to be properly responsive to evolving management needs. The Panel considers that changes in these institutions are the most important of the international actions required. Among the modifications most essential are better means for generating data upon which management decisions can be based and, probably, a change in notions about the nature of the data required for this purpose.

It is notable that only three of the commissions employ their own professional staff for making investigations. It seems plain to us that the decentralized procedure generally employed (each state conducting its own inquiries, with coordination and even some planning on the international level) is not working well. It is possible that improvement in national bodies might alleviate this difficulty, particularly if the funding provided were satisfactory, as it now is not. But perhaps greater progress could be made if the fishery bodies themselves were endowed with an adequately supported, independent staff.

The size and mobility of the fishing fleets working throughout much of the ocean today creates difficulty for regulatory efforts that must be based upon scientific data that can be developed only over a period of years. One avenue of approach is to consider reducing the quantum of data necessary for a decision about the initiation of regulation. The burden of proof for the need for regulation might be erased in this fashion. In a multistate fishery, this course is more easily advocated than accomplished, but exploration of such a solution is warranted by the growing magnitude of the problem. So quickly do fisheries develop to the point of serious encroachment on a stock that it is now, perhaps, wiser to err on the side of premature regulation than to act too late.

Another major improvement that is required is to provide a better match between membership in the fishery commissions of an area and the pattern of fishing there. The far north Pacific and eastern north Pacific provide good illustration. The major fishing states in this enormous area are Japan, the Soviet Union, Canada, and the United States, with Korea and Taiwan emerging. Several fishery commissions have responsibilities in this region, yet not one of them embraces all of these states as members. It is quite difficult to expect improved fishery regulation unless those nations especially active therein join together deliberately to regulate their efforts.

It seems likely that other improvements are also required, but the situation requires far more intensive investigation than the Panel can undertake for this report. The Panel therefore recommends that *a comprehensive study of the nature and effectiveness of present and future arrangements for international fishery regulation to be made by a nongovernmental group. This study should include such problems as methods for obtaining the necessary scientific data, scope of power vested in international fishery bodies, breadth of membership of these bodies, utility of bilateral agreements, and potential usefulness of a global regulatory agency. The results of this study should be used by the State Department as a basis for U.S. ocean policy.*

Available international mechanisms should also be activated to deal with certain other problems. Urgent international action is warranted to re-establish and maintain freedom of scientific research at sea and, more generally, to foster development of a sounder, more effective institutional structure for support of international marine science research. Recent steps in this direction on the international level are encouraging, but further measures are needed and should be made matters of high priority within nations. The critical nature of marine research for resolving problems important to most nations is still not widely understood or appreciated, hence emphasis on this as a priority seems wholly justified. A later chapter in this report is devoted to this problem.

The Panel does not believe that other ocean issues pose policy problems of special urgency at present. It would be convenient if states could act together to settle the several boundary issues that remain unresolved, but, at this time, the kind of agreement that seems feasible would probably not serve the genuine common interests of states. It would be worthwhile if states could stabilize the territorial sea and fisheries limits, but only if the limits adopted were not excessive. It is highly probable that an unsuccessful effort to conclude agreements fixing these limits will spark another series of unilateral extensions even more exaggerated than those now extant. Accordingly, if an international conference is convened to negotiate these issues, it is of the highest importance that it succeed in reaching acceptable solutions.

The Panel also does not consider that the agitation for immediate agreement on the shelf limit and the regime beyond is well founded in terms of imminent prospects for exploitation. These questions deserve the most careful and serious investigation, and international action for this purpose is certainly warranted. However, nothing in the near future in terms of resource exploitation suggests that these problems need immediate solution. It is more necessary for intensified scientific investigation to provide a basis of factual information for ultimate regulation. At the same time, measures must be taken to prevent states from acting unilaterally to extend their control beyond the limits that may be reasonably derived from the 1958 Continental Shelf Convention.

CHAPTER 5

Facilitation of Marine Science Research

TASKS FOR INTERNATIONAL ACTION

The two tasks of principal concern are provision of arrangements securing freedom of access to ocean regions for investigative purposes and logistic purposes, and protection of the platforms and instruments used in research, particularly the unmanned types.

FREEDOM OF ACCESS TO OCEAN REGIONS FOR INVESTIGATIVE PURPOSES

Among those states most active in the conduct of marine science research, and perhaps among an even wider group of states, it is generally accepted that a fundamental task requiring international action is that of removing, or at least attenuating, the numerous restraints on research that arise from extensions of state jurisdiction to ocean regions, both surface and submarine. Until relatively recent times, scientific investigations of the ocean proceeded unhampered. Since 1960 this situation has changed so abruptly and drastically that scientists from several states now confront barriers, impediments, and frustrations never previously encountered. It provides little solace to recognize that the growth of restraints on the conduct of oceanic re-

61

search and data acquisition coincides with the increase in attention both to marine resources, living and otherwise, and to sophisticated military uses of the ocean environment. Whatever the cause, the detrimental impact of these restraints has occasioned sufficient expressions of national and international concern that very few doubt the existence of a problem requiring resolution. For this reason, only the following brief summary identifying legal restraints on research seems required for present purposes.

The most sweeping restrictions on research are those pertaining to waters within state territory: All scientific research within either internal waters or territorial sea requires the consent of the coastal state, if that state so wishes. Less comprehensive controls pertain in the regions beyond, but adjacent to, state territory, and the difficulty of determining the scope of such controls adds to the stringency of the constraint imposed. In the exclusive fishery zones now commonly established by coastal states, fishery research by noncoastal states is regarded as a prohibited activity, but there is uncertainty about what is embraced by "research." Often, scientific research on the continental shelf also must be preceded by a request for permission from the proper coastal state. In all these instances, the relevant boundaries are variously defined or left undelimited by coastal states, leaving the scientist uncertain about the necessity for consent.

In practice, the cumulative impact of this legal framework is to impose substantial restraints on or interference with research. Among the detrimental effects are the wasteful diversion of time and resources from application to productive research, deterrence of entire research cruises and of individual projects, politically induced changes in the nature, scope, and methods of research, and outright refusals of permission for research projects.

A number of factors in combination dictate that the elimination or reduction of these undue restraints on marine science research is attainable only by the concerted action of states. The sheer size of the ocean as the largest surface feature of the planet requires large-scale research efforts extending to every part of the globe. In contrast, political authority over the ocean is highly fragmented and dispersed, being exercised by 103 different states over ocean regions of varying sizes adjacent to them. By necessity, therefore, the states affected or concerned must seek international action, involving at the very least an application to the coastal state for permission to conduct research and, perhaps in specific instances, negotiations over the conditions of access for this purpose.

FREEDOM OF ACCESS TO OCEAN REGIONS FOR LOGISTIC PURPOSES

Experience has demonstrated the need for intergovernmental agreements by which the necessary logistical support for research vessels may be easily available through port facilities in the regions proximate to the expedition. If LEPOR, including the International Decade of Ocean Exploration, involves multistate, multivessel, long-term cruises, as well as smaller and shorter projects, this type of international action will become even more critical for success in attaining the scientific goals of the expanded program.

PROTECTION OF PLATFORMS AND INSTRUMENTS

Recent technological development of new types of vehicles, platforms, and instruments raises problems of their protection against harm or intentional interference. As research submersibles come into increasingly common use over longer operating ranges and buoys proliferate and become larger and endowed with more sophisticated equipment, it is to be expected that regulatory and protective measures will become increasingly necessary to assure their effective use. International action is the only practicable means of achieving the needed regulation and protection.

MECHANISMS FOR INTERNATIONAL ACTION

FREEDOM OF ACCESS TO OCEAN REGIONS FOR INVESTIGATIVE PURPOSES

The goal of marine scientists is maximum freedom of access for scientific investigation and exploration of all ocean regions, irrespective of political boundaries. The extent to which this goal can be approximated may be expected to vary according to the kind of international action sought. Action involving numerous states acting together to prescribe regulations for facilitating scientific research is likely to result in less protection for science than is obtainable in more restricted groupings. In such context, the tendency to settle for the least common denominator in order to assure agreement will probably diminish safeguards for legitimate research work. In every instance the objective should be to secure the greatest protection feasible under the circumstances.

The Panel believes that, on the international level, the principal need is for the conclusion of agreements by which the states concerned either remove handicaps on research or establish a structure for decisions to aid in such elimination. The resulting right of scientists to investigate freely should be coupled with a responsibility to make their results fully available.

In terms of formal international arrangements, a number of alternative means for remedial action seem obviously available: bilateral agreements, arrangements between states in a particular region, the establishment of an international mechanism in aid of individual states seeking, or granting, clearances for areas over which jurisdiction is exercised, and international agreement dealing with the numerous legal issues arising from marine science research in the world ocean. Some believe that these alternatives are not all mutually consistent and that, in particular, to pursue an international procedure, utilizing an international agency, to ease the vessel-clearance problem may inhibit the conclusion of bilateral arrangements to the same end. The Panel believes, to the contrary, that all avenues deserve to be explored and used to the extent feasible and productive of the result of removing, or easing, restraints. At the present time, a preference for any one method, is difficult to justify, since we discern no evidence thus far that conclusion of multilateral agreements will either preclude or discourage agreements between two states or among several states on a regional basis. Contrariwise, nothing has been adduced to indicate that the process of seeking bilateral or regional agreements will interfere with securing a more general multilateral accord. It seems more likely that the process of seeking both types of agreements will be mutually reinforcing, each contributing to the strength and effectiveness of the other.

Bilateral Agreements   The principal need is the conclusion of agreements by which the states concerned either remove handicaps on research or establish a structure for decision to aid in such elimination. Agreements can be either bilateral or multilateral, but, since there is some urgency to remove restraints, conclusion of bilateral agreements may be most desirable initially because of the length of time necessarily involved in achieving more inclusive understandings. A further advantage of early bilateral arrangements is that of acquiring the knowledge and experience that assist in dispelling the suspicion and distrust that sometimes hamper broader agreement.

IOC Arrangements One limited international device for facilitating research consists of the creation of a mechanism for simplifying the task of getting consent for research in areas over which a coastal state exercises jurisdiction. It is hoped that just such a mechanism now exists in the IOC Secretariat in its new role established by the IOC Resolution vi/13 during its 6th Session, entitled "Promoting Fundamental Scientific Research." In accordance with this Resolution, a state using this procedure provides the coastal state and the Commission with a "formal description of the nature and location of the research programme" and

The Secretary of the Commission shall transmit the formal description so received to the coastal state within twenty days of receipt together with the Commission's request for favourable consideration and, if possible, with a factual description of the international scientific interest in the subject prepared by the requesting state, supplemented, if he considers this desirable, by the Secretary.[14]

The primary object of the above terminology appears to be to establish a means for certifying the bona fides of a particular investigation, without simultaneously imposing a great (and probably impossible) burden on the IOC Secretariat by requiring it to engage in an evaluation of each research request sent to it. In this sense the IOC Resolution satisfies, and even improves upon, the recommendation of the U.S. National Committee to SCOR that ". . . the most useful role for the Intergovernmental Oceanographic Commission in facilitating clearances for research vessels undertaking fundamental scientific research would be passive in nature."[15] The U.S. National Committee spelled out what it meant by passive in suggesting that "the IOC upon receipt of requests from member states for research clearances, would immediately transmit them to the concerned coastal state, certifying (when such is found to be the case) that statements are included in compliance with" enumerated conditions concerning handling of data and samples, publication of results, and participation in the research by the coastal state.[16] The IOC Resolution seems to be more satisfactory than the U.S. recommendation since it calls for an automatic favorable recommendation but justifies that by requiring the requesting state to submit a statement of the international scientific interest in the research program described. This statement, supplemented by the IOC Secretary if he thinks it necessary, is then quickly transmitted to the coastal state, thus providing an additional element of support for the bona fide scientific nature of the proposed program.

Although not all requests for clearances will go through the IOC (nor need they unless the coastal state so demands), it is still important that research operators comply carefully with the requirements set out in Resolution vi/13. If scientists are to succeed in avoiding either additional restrictions imposed by coastal states or more onerous administrative difficulties in processing clearance requests both within the requesting state's own government and within an international agency, it seems necessary that they make special efforts to assure that their research activities within areas subject to coastal jurisdiction are executed only for legitimate purposes as previously announced.

The IOC clearance procedure now available to states wishing assistance has other elements that call for some comment. The IOC Resolution now calls for a two-step notification, one very early and preliminary, designed to facilitate planning for participation in the research program by scientists of the coastal state, and a second designed both to elicit a response consenting to the proposal and to permit effective participation by coastal scientists. In sum, the emphasis has shifted from involving the coastal state primarily for securing the necessary consent, with secondary importance attached to scientific cooperation, to facilitating a genuine participation and involvement by coastal scientists in all or some of the proposed program of research. The evidence for this shift consists of the requirement for informing the coastal state as soon as a tentative decision is made to carry out a research program and the omission, in this instance and in the later detailed notice requirement, of any specific timetable for advance notice. It seems reasonable to infer, however, that the preliminary information should be forwarded at least six months prior to the cruise, and preferably earlier, and that the formal description should be forwarded at least 60 days in advance. Compliance with such a schedule should facilitate clearances by giving coastal authorities ample time both to permit planning for participation if desired and to allow officials to check as needed into the nature of the proposed program.

That this timetable requiring early notice will introduce difficulties for some vessel operators is not inconceivable. Two comments are pertinent in this regard. First, it may well be that cruise planners can, by devoting particular attention to the matter, significantly accelerate the time at which a tentative decision on a research group can be communicated to affected coastal states with some assurance that the program will be executed. It is also conceivable that some complaints by scientists about early notice requirements are not wholly justifiable and that the additional attention to this matter is not at all an unreasonable request. Second, even if some inconvenience or added burden of administration results from these notice provisions in the IOC Resolution, this is a very small price to pay for securing a clearance. It is not at all improbable that no consent would be forthcoming unless some such notice provisions were observed.

Another significant feature of the IOC Resolution is that of dealing with the question of handling data and samples between the investigators and the coastal state. Initially, of course, it is the researcher who obtains the data and sample and who uses them for the purposes of the inquiry being undertaken. But in recognition of the coastal interest in the materials acquired during the investigation, including the data about the environment and samples from it, the investigator must make these available to the coastal states. In instances of data or samples that can be replicated, the matter is merely one of the timing of furnishing copies. For items that cannot be duplicated, the Resolution anticipates that they normally will remain in the hands of the investigator but will be made available to the coastal state. This arrangement seems to flow rather naturally from the fact that data and samples are normally in the possession of the investigator who acquires and examines them and from the provision declaring that "special arrangements shall be made regarding the custody of data and samples not feasible to duplicate. . . ."[17] The investigator would retain possession or would retain ownership, in the event that custody is permitted to the coastal state in accordance with a special arrangement. These technical details are of secondary importance to the need for support of scientific research by whoever is carrying it out.

One question about the Resolution deserving brief mention concerns a definition of "fundamental scientific research." It is evident that the Resolution does not attempt to clarify this expression except indirectly, in the eighth paragraph, which states that "the Commission should assist in promoting fundamental scientific research that is carried out either in the framework of the Long-Term and Expanded Programme of Oceanic Research or within Declared National Programmes."[18] At least for the purpose of this Resolution, the IOC procedures apply only to such fundamental scientific research as falls within the two programs. The principal significance of this is the reference to Declared National Programs (since LEPOR is not yet in operation). The designation of a research project as a DNP is the task of officials within each state. Until this Resolution was formulated, such declarations indicated that data from the research would be deposited in World Data Centers and thereby made available to the world scientific community. Now, in addition, a DNP qualifies for IOC asstance with the clearance process.

Quite plainly, the IOC Resolution calls for minimal but meaningful involvement by a central international agency and is thus only a small step away from the normal route of direct state-to-state interaction. As international institutions evolve, growing in experience, capability, and depth of resources and skill, it will be worth a new appraisal to determine whether a more positive role might better facilitate marine science research.

Multilateral Agreements   Less inclusive arrangements than those provided for by the IOC Resolution should also be useful, such as measures applicable on a regional basis or among a limited number of states. Illustrative of the possibilities is the following proposal, which was formulated at the Statutory Meeting in 1967 of the International Council for the Exploration of the Seas:

1. The International Council for the Exploration of the Seas will provide a list of research vessels of the member countries, regularly engaged in scientific investigations. The list will contain such data for each vessel that are needed for identification.

2. Annual cruise programmes will be exchanged between member countries, with the understanding that any member country is free to require a change to be made in a proposed programme of work on its Continental Shelf, if it so wishes. The cruise programmes will indicate as far as possible, where they will impinge on the Continental Shelf, and mention specifically any proposed research on the seafloor.

3. On the basis of the List of Research Vessels and the Cruise Programmes, the member countries are prepared to give, through national office, or agency which they will authorize to act on their behalf, general permission in cases of routine scientific sampling and other probing of the seabed and subsoil and of the bottom fauna by means of grabs and dredges and similar devices.

4. In the case of seismic tests and research involving the use of seismic charge, specific application to undertake such research will continue to be required to each case, and such research will always be dependent upon prior permission.

5. This Agreement is without prejudice to the provision of the Article 5(8) of the Geneva Convention on the Continental Shelf, 1958, and it is on the understanding that recourse may be had to a stricter interpretation at any time.

6. Copies of Cruise Programmes and the general permissions will be deposited in the office of the General Secretary of the International Council for the Exploration of the Sea.[19]

Prospects for successful alleviation of difficulties through means such as the ICES agreement are understandably brighter than for broad international agreements, since unique problems may be dealt with better in the narrower context presented. Since fewer and less diverse participants will be governed by the agreement, it seems more feasible to arrive at general prescriptions favoring free access for research. It may even be possible to establish more expeditious procedures on a bilateral or multilateral basis than might pertain where a potentially large number of states are concerned. In sum, there are marked advantages potentially obtainable through the bilateral and multilateral agreements.

International Convention for Protection of Freedom of Scientific Research Whatever the degree of compatibility among these various methods or approaches, it would not be at all surprising if states considered that their interests are protected best by different methods. Some may be willing to accept the assurances of a clearance procedure established on a multilateral basis and implemented by a centralized agency. Others may prefer to alleviate the difficulties of research and secure the protection of assumed coastal interest by making more selective arrangements with one or only a few other states, that might incorporate provisions that could not practically be included in a general multilateral agreement. Still other states may find it preferable to enter into both kinds of agreements.

In assessing the usefulness of each method mentioned above as a means of removing restrictions on research, the conclusion of a comprehensive international agreement on the legal problems of scientific research is quite plainly a long-term project. We agree with the recommendation of the Stratton Commission on Marine Science, Engineering and Resources that a new convention is desirable to "provide a solid foundation for the freedom of scientists to explore the world's oceans,"[20] and we support the provisions the Commission suggests for inclusion in this agreement. As valuable as such an agreement might be, it would be a mistake to assume that its protection would be sufficient to provide the "solid foundation" so obviously required. It is doubtful, at least at present, that the principles the Commission proposes for this agreement will find easy acceptance by states generally, and particularly by the developing states. The prospect is likely, therefore, that parties to such a general agreement on scientific research will be predominantly those states with a capability for undertaking such research, with far less involvement by the many states without this capability off whose coasts scientific work is required.

Another caveat in regard to the proposed new convention is that new arrangements for regulating other aspects of ocean use may result in new, or unanticipated, restrictions on research. Hence, if it is not practicable to take account of these new problems in a general convention on research, it will be necessary to assure that the interests of science are adequately represented at any conferences dealing with other matters. In short, scientists should continue to express their opposition to impediments that might be derived from regulations aimed at other uses of the ocean and should seek to have remedial or preventive action taken at the meetings that propose such regulations and in the documents that contain them. For the above reasons we do not believe it would be wise to rely completely on a single international agreement (no matter how comprehensive it may be) for protection of science.

The Panel believes a general convention on scientific research is a desirable objective, but it is not likely to provide an overall solution to the problem of removing impediments to research. Scientists should continue to oppose restraints on marine science that derive from agreements directed at other uses of the sea.

FREEDOM OF ACCESS TO OCEAN REGIONS FOR LOGISTIC PURPOSES

It seems likely that the IOC could be usefully employed to expedite entry into coastal waters for various purposes, such as refueling, taking on supplies, picking up equipment and personnel, and recreation. These services are either essential or highly desirable on most scientific expeditions, and there is little doubt that an international mechanism might provide a means for making this type of entry less of a problem. The recently adopted IOC Resolution "Promoting Fundamental Scientific Research" does not provide such a means and does little more than suggest that IOC members recognize that this is a problem. In this Resolution, the IOC "invites interested member states to act in a spirit of international cooperation, to consider favorably and to facilitate within the framework of national laws and regulations the requests for vessels conducting fundamental scientific research to make port calls."[17] In comparison to the provisions that had been recommended to the IOC by its Working Group on Legal Questions Related to Scientific Investigation of the Ocean, which had been detailed and explicit in calling for entry on reasonably short notice, this provision is plainly very light stuff and represents little, if any, progress.

The obvious opportunity for an IOC contribution to this problem is in connection with LEPOR and the expected large-scale cooperative expeditions in implementation of this program. Valuable experience was gained during the International Indian Ocean Expedition, when participating ships and materials were specially marked, and special port facilities were offered by several countries. But perhaps additional measures should be explored, such as a listing of research vessels circulated each year that the program continues. The available alternatives should clearly be examined within the IOC as LEPOR planning proceeds.

PROTECTION OF PLATFORMS AND INSTRUMENTS

Again, various types of international action will probably be required, the precise timing of such action and the nature of the measures involved being dependent on the rapidity with which the newer instrumentalities are used on an international basis and the magnitude of use. At some point, it will probably be desirable, if not essential, to conclude international agreements making provision for the free (or conditional) employment of buoys and submersibles in certain areas subject to the control of an adjacent coastal state. It is plainly necessary that such agreements include provisions for distinguishing scientific research buoys from navigation buoys, where this is necessary, as well as principles designed to protect buoys from harm and to govern the consequences arising from infliction of damage by, or to, research buoys. International arrangements may also be required to make allocations of jurisdiction by which established principles may be applied in concrete situations.

The vastness of the ocean and the attendant difficulty of observing events occurring on or under it might also necessitate the creation of intergovernmental mechanisms by which communications may be routinized and hastened. Reporting of damage to or loss of buoys and of the precise location of research submersible operations might be most effectively facilitated by creating special international institutions, probably as part of an existing body, charged with special responsibilities for these instrumentalities. Where safety of vehicle operation is involved, in addition to efficient use, greater urgency obviously attaches to the invention of more effective institutional means.

At its first session in October 1961, the IOC adopted a resolution on fixed stations that included the recommendation "that steps be taken in consultation with IMCO to clarify the legal status of un-manned and manned observing buoys."[21] The IOC and IMCO have made a number of reports. An IOC group of experts has met a num-ber of times on the problems of ocean data acquisition systems, and a draft treaty has been submitted to IOC and member countries for comment.

RECOMMENDATIONS FOR U.S. POLICY

• *The United States should act to secure the maximum freedom of access for scientific exploration and research in all parts of the world ocean. A variety of international mechanisms for removing re-strictions on research should be used as necessary. Revisions of the Continental Shelf Convention to increase restrictions on research should be resisted; preferably, revisions should eliminate the present "consent requirement."*

• *Any regime for resources of the seabed beyond the limits of national jurisdiction should impose minimal interference with scien-tific research. The United States should oppose allocation of exclu-sive rights of exploration and control of exploration of this seabed.*

• *Freedom of scientific exploration and research should be con-sidered an integral part of the doctrine of freedom of the seas. Scien-tists should oppose restraints on marine science that derive from agreements on other uses of the sea.*

The following discussion amplifies these recommendations.

QUESTIONS OF LAW OF THE SEA

There are at least two important situations now identifiable in which it is or will be essential to provide safeguards for marine science so that regulations for a different ocean use do not unduly interfere with research. At present, the most important such instance is the familiar bundle of difficulties presented by the Continental Shelf Convention. (It is also contended by some that the Convention provisions requiring consent for certain shelf research are already a part of the customary international law.

Recent events within the United Nations and elsewhere strongly suggest that in the near future the world community may seek to re-

vise some or all of the four law-of-the-sea conventions adopted at
Geneva in 1958. As a party to all of these treaties, the United States
obviously has an important interest in assuring that such revisions
serve its interests and those of the world community generally. Although scientific research is or may be affected by each of these agreements, the Panel is now most concerned about U.S. policy toward
revision of two of the treaties, the Continental Shelf Treaty and High
Seas Treaty. Policy in regard to the latter is also important because it
appears similarly likely that a new international treaty will soon be
negotiated to provide a legal regime for regulating nonliving-resource
development in areas of the high seas beyond national jurisdiction.

REVISION OF THE CONTINENTAL SHELF CONVENTION

It should be the policy of the United States to eliminate or to substantially minimize the impediments to scientific research established for
the first time by the 1958 Continental Shelf Convention. Deletion of
Article 5(8) would obviously contribute to this end but would not be
sufficient. In providing that each coastal state has exclusive rights of
exploration of its shelf, it is reasonably plain that, even without Article
5(8), some (perhaps many) coastal states would forbid even bona fide
scientific research on their shelves on the ground that it was inconsistent with that state's *exclusive* right to explore, which (it would be
claimed) is indistinguishable from scientific research. Accordingly, the
only certain way of freeing scientific research is to abolish the coastal
state's exclusive right of exploration. Unfortunately, it does not seem
likely that coastal states are prepared to relinquish their newly acquired
sovereign rights of exploring and exploiting the natural resources of
the shelf. The question thus is how to acknowledge this right but to
minimize its impact on bona fide scientific research.

One desirable alternative would be to secure the deletion of Article
5(8), with its express requirement of consent, and to substitute a requirement for notice of certain intended research. The coastal state
would continue to be authorized to refuse to allow the research, but
voiced or written objection would be required. In the absence of objection within a stated period after timely notice, the particular research would be considered authorized. It may be helpful in this connection to insert a provision in the revised Convention recognizing the
importance and value of freely conducted scientific research regarding
the shelf.

Another but less attractive alternative is to maintain the present formulation of Article 5(8), while continuing to improve the procedures by which clearances for research are obtained. As experience develops with the use of the recently adopted IOC procedure, ways for strengthening this avenue in a manner that would improve it should be explored.

REVISION OF THE HIGH SEAS CONVENTION

It is well known that the Convention on the High Seas omits mention of "freedom of scientific research" as one of the freedoms expressly embraced by "freedom of the seas" as defined in Article 2 of the Convention. If revision of this treaty is on the agenda of a new law-of-the-sea conference, it should be U.S. policy to revise the definition of freedom of the seas to include the conduct of scientific exploration and investigation. Recent indications are that developing states are unconvinced that research enjoys the same freedom as, for example, navigation, and it would be beneficial to record the widespread support for freedom of research by including a specific provision to that effect in the High Seas Convention.

Such a revision of this treaty, wholly desirable in itself, has considerable added significance because of the anticipated conclusion of still another treaty dealing with the high seas region, namely an agreement on a legal regime for development of the nonliving resources of the seabed underlying the high seas beyond the continental shelf. As noted earlier, it is to be expected that allocation of exploration rights will require an accommodation with scientific research, and vice versa.

The Panel believes that it is essential to structure the regime for exploration and exploitation of nonliving resources beyond the limits of national jurisdiction in a fashion that poses the absolute minimum interference with or restraint upon scientific research. The Panel also believes it should be the policy of the United States to include freedom of scientific exploration as an integral component of the doctrine of freedom of the seas.

One kind of problem to be confronted is illustrated by the proposal of the Commission on Marine Sciences, Engineering, and Resources in its Report to the President in January 1969. The Commission makes numerous recommendations concerning a regime for the deep seabed, among which is the creation of an International Registry Authority that will register claims to explore for particular minerals in a particular area. Such registry, open only to the first to appear as long as he is technically and financially competent and willing to undertake exploration, confers an exclusive right of exploration. The Commission says, however, that "preliminary investigation" is free so that prospective claimants may "determine whether it is worthwhile to register a claim to explore."[22] The distinction between "preliminary investigation" and "exploration" is not explained, and no procedure is set out for resolving disputes on the point. One observer comments on the resulting dilemma:

It is impossible to ascertain, therefore, whether one may engage in preliminary investigation of an area where another has been given an exclusive claim to explore and/or to exploit. This uncertaintly would lead to misunderstanding and could have a stifling effect on the freedom of the seas for both commercial and scientific inquiry.[23]

Perhaps a more serious problem implicit in the Commission proposal is attendant upon the grant of an exclusive right to explore. If such a right exists, it would follow either that scientific exploration is no longer permissible or that some procedure must be established to assure that a given project *is* scientific in nature and not intended to conflict with the commercial enterprise. As scientists well know, the latter alternative poses very considerable problems for them in terms of administrative and political difficulties.

The Panel believes that neither of the above problems should arise since we perceive no need for creating any exclusive rights of exploration in the deep seabed nor for a system to control such exploration. Such rights are wholly unnecessary for this area in the foreseeable future. Previous experience in exploration of the continental shelf area and the knowledge already accumulated about surficial deposits in the deep sea demonstrate that oil and mining interests do not require exclusive rights to induce them to explore. Mr. George Miron, Washington, D.C., attorney, summarizes the situation:

. . . an oil and gas prospector needs tenure before proceeding to invest in drilling but does not need tenure in order to do geophysical exploratory work that does not require emplacement of permanent structures on the ocean floor. Accordingly, those seeking to explore for oil and gas by geophysical techniques should be free to do so on a non-exclusive basis. Similarly there is no reason to assure tenure to those who would explore surficial deposits since such work has been and can be done on a non-exclusive basis without assurances of tenure.[24]

While the Panel believes that both the above problems can be avoided by the exercise of some realism about requirements for deep-sea mineral exploitation, it is distinctly possible that some system for

accommodating various, possibly conflicting, rights of users is highly
desirable and, perhaps, necessary. Such a system would be built on the
premise that the conduct of research is recognized as an exercise of
freedom of the seas and is protected by the latter principle.

PROTECTION OF PLATFORMS AND INSTRUMENTS

Consideration of measures for protection of ocean data acquisition
systems has occurred thus far wholly within international technical
bodies such as the IOC and IMCO. Progress to date on concluding protective arrangements is commendable, and the United States should
continue to support this approach. At the same time, other United
Nations bodies, especially the General Assembly and its subsidiary
organs and committees, have devoted increasing attention to issues of
ocean regulation that present political, legal, and economic problems
of a substantial nature. It appears very likely, at the present writing,
that these many issues will be the focus of the attention of very large
multilateral gatherings over the next several years. It is not to be expected that this process of consideration and decision will produce
decisions quickly, since the problems are complex and negotiations
understandably deliberate.

In this context it is highly desirable to continue consideration of
ODAS protection in the technical groups in which the process began.
If this matter were subsumed into the general negotiations on law-of-
the-sea issues, the result would probably be to politicize issues that are
primarily technical in nature and to delay adoption of suitable arrangements for several years. Neither of these effects is desirable.

*The United States should actively support a treaty to protect un-
manned buoys, free-floating instrument packages, and other ocean
data acquisition systems now in use and under development. To expedite agreement on an ODAS treaty, the United States should resist
all attempts to tie this question to broader problems of the law of the
sea, including the limits of national jurisdiction.*

POSSIBLE UNILATERAL ACTION BY THE UNITED STATES

*The United States should make a unilateral declaration allowing sci-
entific research in areas outside internal waters but subject to its jurisdiction, provided that certain conditions are observed.* In order to
assure that the activities of research vessels are, in fact, bona fide sci-

entific activities and that the results of their work will be promptly and
fully available to the United States, the United States should

Be given reasonable advance notice, a period of 60 days probably
being adequate.

Have the opportunity to participate in the research and exploration
and have access to all equipment, compartments, and instruments
aboard the vessel.

Have the right to receive copies of all data on request, and the right
of access, for study, to all samples not feasible to duplicate.

Be assured that significant research and exploratory results will be
published in the open scientific literature.

Be assured that the scientific exploratory activities will present no
hazard to the resources or uses of the sea or seabed (e.g., seismic ex-
plorations that could damage fish stocks, or exploratory drilling that
could result in petroleum pollution).

If these conditions are prescribed and advantage taken of them as may
seem desirable to the United States, it is believed that U.S. interests in
these areas will be fully safeguarded against injury from any cause.

This proposal for unilateral action was forwarded to the Department of State in May 1970 by the president of the National Academy
of Sciences along with a resolution by the Council of the Academy
urging vigorous exploration by the federal government of ways and
means of encouraging all nations to ease restrictions on marine research. The proposed unilateral action is still under study within the
executive branch at this writing.

It is the Panel's belief that unilateral action substantially similar, but
not necessarily identical, to the above would provide a dramatic demonstration by the United States of its belief in the importance of free
ocean research. A bold unilateral initiative of this kind can be extremely
effective in demonstrating the advantages, both to the United States
and to any other state so acting and to nations generally, of encouraging free and open scientific research and exploration in areas under
their jurisdiction, and it might soon become widely apparent that this
is a very economical means of acquiring information of special value to
the coastal state.

It is particularly fitting that the United States take this initiative.
Our position of world leadership in the actual conduct of ocean sciences should be paralleled by equivalent leadership in general support

of freedom of scientific inquiry and, specifically, in removing or attenuating restraints on such research. In any event, urging developing
nations to take this action seems unlikely to be successful in view of
their deep suspicions of the motives of the more developed nations. A
demonstration by the United States, wholly without regard to any advance commitment of reciprocal action, would appear to be the best
approach.

A number of desirable consequences might reasonably be expected
to follow from the unilateral action recommended. An initial impact
could be to dissipate some of the suspicions recently attached to U.S.
scientific expeditions operating in areas subject to the jurisdiction of
other states. By demonstrating its willingness to enable foreign scientists to operate in its waters subject only to minimum safeguards, the
U.S. might well create a greater trust in its own activities abroad.

The action we recommend could, by itself, induce reciprocal action
by other states around the world, opening up areas now either closed
or subject to burdensome conditions and uncertainties. Our unilateral
action might provide a substantial argument to foreign scientists who
may wish to persuade their governments to reduce national obstructions to foreign research. Most marine scientists, if not their governments, are fully aware of the high value of freedom of inquiry and are
also aware of the very real benefits to be gained from foreign scientists
working in adjacent waters. The information and data thus made available can be a real contribution to the scientific work of coastal scientists.

Even if similar reciprocal action is not induced by this U.S. action,
it is not inconceivable that it could have the effect of encouraging the
conclusion of bilateral agreements between the United States and
other states. Some states may not feel able to reciprocate by action as
liberal as that suggested here and may feel ill-equipped to accept the
exact assurances we would accept. However, given this display of U.S.
goodwill, some states might well believe that, by agreement with the
United States, the coastal state could assure itself of procedures and
principles that satisfactorily meet its interests and requirements and,
at the same time, reduce obstructions to U.S. scientific efforts in its
waters.

A further consideration, more long-run than others, is the realization
that unilateral action of the type described here could bolster the general position that the United States should assume in the future (as it
has in the past) in vigorous support of the concept of freedom of the

seas. It continues to be in the interest of the United States, indeed of the entire world, to maintain the oceans as open to the utmost freedom of use, with restraints imposed only when required for the greater good. Outstanding international legal scholars, such as Professors Mc-Dougal (Yale University Law School), Henkin (Columbia University School of Law), and Auerbach (University of Minnesota Law School), concur in urging that the United States reaffirm its support of freedom of the seas and are specific in urging the widest possible freedom for marine science research. Unilateral action by the United States would unequivocally establish its commitment to freedom of the seas and strengthen the voice and influence of the United States in confronting the coming challenges to that concept.

CHAPTER 6

Assistance to Developing Nations

Because of broad political and economic implications, it is difficult to deal with the question of assistance to developing countries strictly in the context of international marine science affairs. This chapter in the Panel's report differs somewhat from the previous chapters in that it represents only preliminary examination of some of the problems involved in the development of programs of assistance in matters pertaining to the ocean. Although some tentative suggestions are advanced, the major conclusion is that a comprehensive study of the subject is required.

WHY TECHNICAL ASSISTANCE IS REQUIRED

It is not necessary at this stage to consider whether nations will attempt to benefit from resources in the sea adjacent to them. Already a very large number of states, including many with developing economies, make use of these resources or have an interest in doing so. However, to gain full or even substantial partial advantage from the availability of such resources, it will frequently be desirable that the state itself develop the scientific institutions, capabilities, and facilities that will permit adequate investigation of ocean resources and lay a base for their exploitation. The vast majority of countries bordering the ocean lack trained manpower and facilities for this investigation, and an effective means for building strength in this field is through the assistance of the states having competence in it.

In addition to these substantive justifications for assistance programs in marine sciences, there are obvious political reasons for such activities. For the past decade, the political relationships between technologically advanced nations and the developing nations have been undergoing a marked transformation as the latter group becomes increasingly important on the international political scene. Among the several bases for this change, we would note only the abrupt increase in the number of developing nations, which has had a great impact on the balance of voting strength in the United Nations. It is now difficult, if not impossible, to plan for broad international ocean programs, whether concerning research or exploitation, without taking into account the interests of developing nations. In most cases, it is necessary to involve these nations at least to the extent of their normal support and it is often necessary to get their active cooperation. As was discussed before, the consent of these states for access to areas under their jurisdiction is frequently required in order to carry out geological, biological, and ecological research. It is highly desirable to secure the endorsement and cooperation of the developing nations for scientific programs of mutual interest.

SCOPE OF ASSISTANCE PROGRAMS

A major difficulty in arriving at improved methods of providing technical assistance in marine science is in accommodating the interests of developing nations on a matter that is politically sensitive and of greatly different importance to individual developing states. A number of factors bear significantly on the dual problem of marine science research subjects for which assistance to developing states can usefully be extended and the level or degree of sophistication and applicability to development that should be incorporated in an assistance program. A review of some of these factors follows, with consideration of an approach that might answer the dual problem.

ECONOMIC DEVELOPMENT

The governments of developing nations are understandably preoccupied with problems of economic development. Proposals for international ocean programs will receive generally negative, or at best lukewarm and skeptical, responses unless they have direct bearing on resources development, particularly the development of living resources, or unless substantial assistance for this purpose is provided in connection with such programs. Disillusioned, and perhaps rightly, about international assistance and cooperation, the enthusiasm of the leaders of many of these nations has diminished. There have been too many ill-founded assistance schemes that turned out to be costly to the recipient countries. A number of examples could be cited involving both bilateral and multilateral relationships. In light of this background, any proposal for an international ocean program will naturally be scrutinized by officials of developing nations from the point of view of what their countries can get out of it and at what cost.

CHARACTERISTICS OF DEVELOPING STATES

The developing nations cannot be treated as a homogeneous group of states. They are far more heterogeneous than developed nations in standard of living, degree of industrialization, average level of education, availability of technically trained personnel and research facilities, and appreciation of the role of science in resource development and management. Because of such widely varying characteristics, these states differ in their ability to participate in, or benefit from, regional or worldwide programs of scientific investigation. As a result, the question of how to provide effective international assistance to nations of such wide variety has been the most vexing one confronting international aid agencies. This also explains the fact that assistance programs for individual states with specific objectives have generally been more successful than ambitious and often unrealistic regional programs.

INTERRELATIONSHIP OF PROBLEMS IN MARINE AFFAIRS

Because of the nature and complexity of marine affairs and their recent prominence in an international political context, the international problems arising from wholly different fields of ocean activity are becoming more and more inseparable in negotiating situations. Even when actual physical interactions between disparate activities are minimal, the activities tend to be considered together because of this close legal and political relationship. Thus, although it still makes scientific sense in most cases to handle international fishery management problems as distinct from those of mineral resources, discussion of the latter tends also to bring in the former. Almost any discussion of jurisdiction over the seabed has some implication for the question of jurisdiction over superjacent waters. It is particularly relevant to the present discussion that many nations (and not only developing nations) believe that much scientific research in the ocean has commercial implications. Ocean research is, thus, inseparable from exploitation in the thinking of many nations, particularly those classified as developing.

CONSTRAINTS ON INTERNATIONAL AID

No matter how we appraise the results of assistance programs under different auspices, the proportion of successful projects is rather low. Except for institutions providing only relatively hard loans, such as the International Bank for Reconstruction and Development, international aid organizations are expected to assist as many nations as possible. Instead of concentrating their efforts where the probability of success is highest, they have to distribute their resources even among the countries where the odds are obviously against them. For example, fishery projects supported by the United Nations Development Program have often been criticized on the ground that funds are spent in areas where resource potentials are not great. But even to nations with limited fishery resources in adjacent waters, fishery development may still be important, particularly if, as is often the case, other possibilities are equally or even more limited. The administrator of the UNDP cannot say to these nations that he will not recommend projects for them since the same amount of money would result in much greater returns somewhere else. On the other hand, the administrator has the responsibility of using international funds most effectively and should, in this case, have an overall understanding of long-term development potentials of different areas of the world ocean as well as of the ability of different nations to realize such potentials. The projects actually recommended and approved reflect these two considerations.

In sum, the factors outlined above make it difficult to establish programs in the broad area of marine science that effectively provide assistance to developing states. One means of coping with these problems might be to formulate an integrated program that extends coverage horizontally to a number of priority fields, or subject areas, of marine science and vertically incorporates different levels of research, education, survey, and development activity. By this means, the assistance plan might be made broad and varied enough to permit states of greatly varying capability and interest to participate in ways that might be valuable to their own interests.

It is not to be expected that such a comprehensive program should or could be implemented, or even coordinated, under the auspices of one international agency. But preparation of plans for such a program would entail indication of the positions of specific fields (or subject matters) in terms of the priority needs in ocean investigation, thus providing a form of guidance that might otherwise come from an overall planning agency. On the vertical axis, the proposed program would incorporate activities at various levels of sophistication and applicability to development problems, so that nations at greatly different levels of development could identify their appropriate participation in the program. These would vary from the very practical to the more abstract long-range research program. Developing countries would, if past experience is a guide, be principally interested in increasing their exploitation of living resources and, perhaps, of petroleum and mineral resources. Most, if not all, of their effort would be directed to problems of immediate applicability. It would be especially useful to have expert assistance available for use by individual states desiring guidance in determining the appropriate level of activity in which they should engage.

For illustration of this approach, the Ponza Report[9] offers useful clues. The report identifies several priority subject fields for investigation: ocean circulation and ocean–atmosphere interaction, life in the ocean, marine pollution, and dynamics of the ocean floor. Within each of these very broad and general categories the Ponza group further identified (albeit tentatively) more specific projects that might be implemented in the long-term and expanded program of ocean research and exploration. For example, six projects are proposed within the general area of life *in* the ocean: primary and secondary production measurements, exploratory surveys (with eight geographical areas to be given priority), living resources of the Antarctic seas, detailed ecosystem studies, development of coastal aquiculture and improvement of its scientific base, and the stock and recruitment problem. This list indicates that some aspects of the program suggested by the Ponza group approach the level of development activities that might appeal to a wide range of states. The list includes, for example, "exploratory surveys" and "coast aquaculture," in which a number of the developing nations would be willing to participate and are capable of doing so. Under "dynamics of the ocean floor," projects that might interest some of the developing countries include "geological and geophysical surveys of continental margins" and "river mouth monitoring," both undertakings having substantial practical implications. Specific projects are also suggested in connection with marine pollution, a subject of practical importance to everyone.

The Ponza Report did not systematically consider the matter of projects suitable at different levels of economic development, beyond the manner already indicated above. But the group was requested by the IOC Bureau and Consultative Council to consider two closely related questions: "How can ocean exploration and research best contribute to the particular needs of the developing nations? and "How can increased ocean research activities by the developing countries contribute to their social and economic development?" The group did not really attempt to respond to these questions except for the general notation that participation by such states in research provides an information base and a skilled group required for resource development. The developing states demand more than this level of generality; their interests are primarily in specific ocean-oriented development projects and in the amount and kinds of assistance the developed nations can and will provide them.

LONG-TERM EMPHASIS ON ASSISTANCE PROGRAMS

However concerned the developing nations may be about economic development aspects of assistance, the scope of assistance in the context of international marine science affairs must go beyond this to emphasize development of scientific competence and educational institutions. All available channels of assistance (multilateral or bilateral, governmental or nongovernmental) should be explored to achieve this objective. In this area, too, the widely varying characteristics of developing countries make it difficult to apply standard approaches. Full analysis of experiences, obtained under various assistance programs for developing research or educational institutions in oceanography, meteorology, and fishery, should be undertaken. It will be appropriate to hold a meeting of scientists and administrators experienced in this field to develop guidelines for future programs.

The academic communities of the technologically advanced nations should be prepared to make their services available to assistance programs in marine sciences for developing countries. Fellowships are provided fairly widely for training of marine scientists from developing states. Participation by scientists from advanced nations in training and education within developing states is still limited.

Experience over the last two decades indicates the desirability of expanding the role of nongovernmental institutions in this area. Direct association between universities in the developing nations and those in the developed countries has proved particularly effective as a means for building up scientific and technical competence and deserves greatly increased financial support from both public and private sources.

NEED FOR FUTHER STUDY

A full analysis of the question of assistance in marine science affairs is clearly beyond the scope of the present study. Before specific recommendations on future action, either at a national or international level, can be made, *an appropriate component of the National Academy of Sciences should undertake a comprehensive study of the nature, scope, and effectiveness of various attempts to provide technical assistance to developing countries in marine science and resource development and in other aspects of ocean affairs.* In order for such a study to be really useful, it would have to cover a wide variety of subjects, and should include, *inter alia*

Review of past and present programs of assistance, under different arrangements (multilateral and bilateral; governmental and nongovernmental) at various levels ranging from basic oceanography to fishery development and in different geographic areas

Critical appraisal of achievements of these assistance programs

Evaluation of the ability of the countries receiving assistance to participate in or benefit from these programs

Examination of procedures and operating methods of different funding institutions and executing agencies from the point of view of their effectiveness in providing assistance in ocean affairs

Suggestions for more effective institutional arrangements, including the possibility of establishing a special fund for assistance in ocean affairs, and proposals for specific programs of regional or global scope that should receive priorities

Special consideration based on the needs and capabilities of the recipients of assistance in fishery development, which is obviously the most immediate and practical aspect, and training and education in marine science, perhaps the most important aspect in the long run.

ASSISTANCE IN FISHERY DEVELOPMENT

Among various activities concerning the ocean, the largest amount of assistance has been provided in fishery development. Many different arrangements have been employed. At the most practical level, inputs of foreign capital, management, and technology have had tremendous effects on the development of fisheries in underdeveloped areas. At least at initial stages, most of the fishmeal, tuna, and shrimp industries in those areas developed under some arrangements with foreign countries. However, these are give-and-take kinds of arrangements and may not be included in the ordinary concept of "assistance." There have been a great many private or governmental bilateral schemes for grants, loans, or supplier's credits. Some of them have been reasonably successful, but many others have failed miserably. Quite a few have turned out to be equipment-selling projects (often very expensive ones) rather than assistance schemes. The projects of the IBRD which has also been involved in financing fishery (and harbor) schemes, might be considered generally successful; in any case, the IBRD usually gets its money back. The Asian Development Bank is now quite active in fishery assistance as well as in development of oil and gas resources; some of their projects appear to be good ones.

At the level of technical assistance and preinvestment activities, bilateral arrangements (mainly governmental) and UNDP-financed projects, executed by the specialized agencies (in most cases FAO in fisheries), are the two main sources of support. A few of the bilateral projects have succeeded (such as Thai–German and Indo–Norwegian), but many others have failed. Bilateral arrangements have been generally effective for training and education. Direct association, formal and informal, between institutions in developed countries and those in developing countries has been a particularly successful basis of assistance and has contributed towards building up technical and scientific competence in ocean research in the developing nations.

The nature and scope of UNDP-supported fishery projects, described in many other papers, needs no further elaboration. The UNDP represents the largest source of assistance in fishery development at the level of technical assistance and preinvestment activities, in terms of the total amount of money spent as well as the number of technical people involved. Although many of the projects have been unsuccessful, performance is generally improving. Projects in some other fields of ocean activities, such as hydrographic surveys, maritime training, feasibility studies for harbor construction, and meteorology, are also supported by the UNDP.

If part of the funds now used for national projects directly oriented toward fishery development were diverted to support broader international ocean research projects with practical implications, such as some of the projects proposed in the Ponza Report, the problem of incorporating assistance aspects in a long-term ocean research program might be partially solved.

The Indian Ocean Fishery Survey and Development Project, the preparatory phase of which has been approved by the UNDP, is, perhaps, the first one of its kind. Although the project included development aspects, the survey part of it coincides with one of the priority projects recommended by the Ponza group under "Life in the Ocean." The UNDP is supporting its preparatory phase with the understanding that some of the developed nations will participate in the survey phase to make it a joint undertaking.

As long as the funds used for schemes of this nature remain relatively small (and the counterpart contributions required are minimal), as in the case of the preparatory phase of the Indian Ocean project, the above approach will not be objected to by the governments of developing nations. It might work, provided that the developed nations are indeed prepared to participate in the program in a substantial way. However, it is not very likely that the majority of developing nations would endorse the idea of diverting to broad ocean research programs a large portion of the total amount of funds now made available to support national development projects. It is not a question of whether these nations would benefit from such programs, for the Indian Ocean Fishery project, if properly planned, adequately financed, and efficiently executed, will undoubtedly benefit all the countries bordering the Indian Ocean, though in varying degrees. It is a question of whether they would be prepared to endorse such a new international endeavor at the expense of support to their own national projects.

In order to develop successful projects of this kind, the UNDP would be required to make substantial changes in its policies and procedures. The present policy of providing support only to the developing nations causes difficulties in formulating regional ocean projects in which all nations having either research or commercial interests in a region can participate. The requirements for matching contributions are too rigid; in many cases it is 50 percent and sometimes even higher. The arrangements for project implementation do not fit in with the needs of broad ocean programs.

While these problems might be overcome in due course, a better alternative would be a separate funding arrangement for the specific purpose of exploring this area of international cooperation. Funds might be channeled through one of the existing international funding institutions, possibly the UNDP, under a fund-in-trust arrangement. The administration of the funds would be left to a small executive group. A team of high-level advisers would assist in overall planning, selection of projects, arrangements for implementation, and evaluation of achievements. A precedent of this sort is found in the operation of the Fund for the Development of Irian (formerly Netherlands New Guinea), a fund-in-trust arrangement between the government of the Netherlands and the UNDP.

The requirements for matching contributions would be kept flexible and would be determined on the basis of each nation's ability to participate in a project and to benefit from it. The executing agency would be a capable international or national institution, such as IOC or FAO.

References

1. United Nations Educational, Scientific and Cultural Organization, 1960. A resolution establishing an Intergovernmental Oceanographic Commission within the United Nations Educational, Scientific and Cultural Organization. 11 C/Resolution 2.31, Article 1, sec. 2.
2. Commission on Marine Science, Engineering and Resources, 1969. Our nation and the sea: A plan for national action. U.S. Government Printing Office, Washington, D.C. 305 p.
3. Scientific Committee on Oceanic Research, 1967. International ocean affairs: A special report prepared by a joint working group appointed by Advisory Committee on Marine Resources Research of the FAO, SCOR, and Advisory Committee of the WMO. La Jolla, California. p. 6–8.
4. Intergovernmental Oceanographic Commission, 1969. Comprehensive outline of the scope of the long-term and expanded programme of oceanic exploration and research. Part II: Practical problems of implementation, para. 5, Supporting services. Paris.
5. Intergovernmental Oceanographic Commission, 1969. Comprehensive outline of the scope of the long-term and expanded programme of oceanic exploration and research. Part II: Practical problems of implementation, para. 7, Integrated global ocean station system (implementation aspects). Paris.
6. Comité International de Geophysique (of the International Council of Scientific Unions), Guide to international data exchange through the World Data Centers. CIG-IGSY Secretariat, London.
7. Commission on Marine Science, Engineering and Resources, 1969. Science and Environment, panel reports, U.S. Government Printing Office, Washington, D.C. Volume 1, p. II-59.
8. Intergovernmental Oceanographic Commission and World Meteorological Organization, 1969. General plan and implementation of IGOSS for phase I. SC/IOC-vi/21. Paris meeting 27 October 1969. Section 1, p. 1–2.

91

9. Joint Working Party of SCOR, ACMRR, WMO, 1969. Glogal Ocean Research, Ponza and Rome, p. 25.

10. Intergovernmental Oceanographic Commission, 1969. Comprehensive outline of the scope of the long-term and expanded programme of oceanic exploration and research, Part II, Practical problems of implementation, para. 2, Data and information management.

11. Intergovernmental Oceanographic Commission, 1969. Comprehensive outline of the scope of the long-term and expanded programme of oceanic exploration and research, Part II, Practical problems of implementation, para. 3, Instrumentation and methods, Paris.

12. Treaties and Other International Acts Series, 1964. Convention on the Territorial Sea and Contiguous Zone, Article 1. TIAS 5639, p. 3

13. U.S. Delegation to the Eleventh Meeting of the Bureau and Consultative Council of the IOC, 1970. Delegation report, Paris 1/16–20/70, p. 47.

14. Intergovernmental Oceanographic Commission, 1969. Resolution vi/13 adopted at the 6th Session of the IOC, Doc SC/IOC-vi/32, 23 September 1969, p. 25–26.

15. Calhoun, J. C., 1969. Letter on behalf of U.S. National Committee to SCOR to Chairman of SCOR, 8 January.

16. Calhoun, J. C., 1969. Letter on behalf of U.S. National Committee to SCOR to Chairman of SCOR, 8 January.

17. Intergovernmental Oceanographic Commission, 1969. Resolution vi/13 adopted at the 6th Session of the IOC, Doc. SC/IOC-vi-32, 23 September 1969, p. 26.

18. Intergovernmental Oceanographic Commission, 1969. Resolution vi/13 adopted at the 6th Session of the IOC, Doc. SC/IOC-vi/32, 23 September 1969, p. 25.

19. Working Group on Legal Questions Related to Scientific Investigations of the Ocean, 1968. Summary report of the 1st meeting, Paris, 16–20 September, 1968, Annex IV. Doc. AVS/9/89m(8).

20. Commission on Marine Science, Engineering and Resources, 1969. Our nation and the sea: A plan for national action. U.S. Government Printing Office, Washington, D.C. p. 203.

21. Intergovernmental Oceanographic Commission, 1962. Report of the first session, Paris, 19–27 October 1961. UNESCO/NS/176. Resolution I-7, operative para III, fixed stations.

22. Commission on Marine Science, Engineering and Resources, 1969. Our nation and the sea: A Plan for national action. U.S. Government Printing Office, Washington, D.C. p. 148.

23. Miron, G., 1970. Proposed regimes for exploration and exploitation of the deep sea-bed, p. 98. In M. Alexander [ed.], Proceedings of the fourth annual conference of the Law of the Sea Institute, June 23–26, 1969. University of Rhode Island, Kingston.

24. Miron, G., 1970. Proposed regimes for exploration and exploitation of the deep sea-bed, P. 108. In M. Alexander [ed], Proceedings of the fourth annual conference of the Law of the Sea Institute, June 23–26, 1969. University of Rhode Island, Kingston.